Handbook of Nutrient Requirements of Finfish

Editor

Robert P. Wilson, Ph.D
Department of Biochemistry and Molecular Biology
Mississippi State University
Mississippi State, Mississippi

CRC Press
Taylor & Francis Group
Boca Raton London New York

CRC Press is an imprint of the
Taylor & Francis Group, an **informa** business

First published 1991 by CRC Press
Taylor & Francis Group
6000 Broken Sound Parkway NW, Suite 300
Boca Raton, FL 33487-2742

Reissued 2018 by CRC Press

© 1991 by Taylor & Francis
CRC Press is an imprint of Taylor & Francis Group, an Informa business

No claim to original U.S. Government works

A Library of Congress record exists under LC control number: 91207170

Publisher's Note
The publisher has gone to great lengths to ensure the quality of this reprint but points out that some imperfections in the original copies may be apparent.

Disclaimer
The publisher has made every effort to trace copyright holders and welcomes correspondence from those they have been unable to contact.

ISBN 13: 978-1-138-10586-7 (hbk)
ISBN 13: 978-1-138-56000-0 (pbk)
ISBN 13: 978-0-203-71214-6 (ebk)

Visit the Taylor & Francis Web site at http://www.taylorandfrancis.com and the
CRC Press Web site at http://www.crcpress.com

PREFACE

The total worldwide aquacultural production of finfish exceeded 7 million metric tons in 1988. This production involved the culture of an estimated 300 different fish species. Finfish are cultured throughout the world by extensive and intensive culture systems. Little to no supplemental feed is used for extensive culture, whereas supplemental feeding is required for all intensive fish culture. Thus, the successful intensive culture of any finfish requires the availability of nutritionally complete feeds that will support good growth of the species being cultured. Some of these intensive cultured species are fed diets of natural foods, such as trash fish, others a mixture of natural food and processed feeds, whereas others are fed only processed feeds. Economic pressures and environmental concerns are forcing more and more culture systems to shift to using nutritionally complete processed feeds. These factors add additional pressure to the fish nutritionist to develop more cost-efficient and low-pollution diets for the aquaculture industry.

Of the estimated 300 different fish species being cultured worldwide, limited nutritional information is available on less than perhaps 50 species. In this book, we have summarized the qualitative and quantitative nutrient requirements for almost all of the cultured finfish for which a significant amount of nutritional knowledge exists. You will note that the nutrient requirement data for certain species, such as the channel catfish and common carp, are nearly complete, whereas these data are just becoming available for several other species.

This book differs from others dealing with fish nutrition in that it summarizes the current nutritional information by species. Each chapter contains an introduction which includes information on how the species is cultured, some index of production, and the regional locations where the species is being cultured. An example of a purified or test diet and any special conditions necessary for laboratory studies on the species are provided in most chapters. Nutrient requirement data are discussed and summarized for each species. An example of a practical diet formulation is provided for most species. A discussion of special diets and feeding practices is included for certain species.

This book should be useful to students, researchers, practicing nutritionists, aquaculturists, and feed manufacturers who are interested in fish nutrition. It should serve as a valuable handbook of our current information on fish nutrition as well as a valuable reference book to obtain specific information about a certain species. It is hoped that the information — and lack of information — contained in this book will stimulate researchers to continue to expand our knowledge in this vital area of nutrient requirements of fish.

The editor gratefully acknowledges the contributions made by the chapter authors. Several of the international contributors have included information from local and regional publications that had not previously been available. In addition, some contributors have been kind enough to share some recent unpublished data with us.

EDITOR

Robert P. Wilson, Ph.D., is Professor of Biochemistry and Molecular Biology at Mississippi State University, Mississippi State, Mississippi. He received his degrees from the University of Missouri, Columbia, obtaining the B.S. Ed. in 1963 and the M.S. and Ph.D. in Biochemistry in 1965 and 1968, respectively. He served as an Instructor of Agricultural Chemistry at the University of Missouri until 1969 and as Assistant Professor, Associate Professor, and Professor in the Department of Biochemistry at Mississippi State University from 1969 to 1989. Dr. Wilson also served as Head of the Department of Biochemistry from 1979 to 1989. He spent six months as a Visiting Professor at the Institute of Marine Biochemistry in Aberdeen, Scotland, in 1984. He assumed his current position as Professor of Biochemistry and Molecular Biology in 1989.

Dr. Wilson is a member of the American Institute of Nutrition, Catfish Farmers of America, Catfish Farmers of Mississippi, and World Aquaculture Society. He has served as a member of the Subcommittee on Nutrient Requirements of Warmwater Fish of the Committee on Animal Nutrition, National Research Council, National Academy of Science on three separate occasions. He has served as the U.S. representative on the Fish Nutrition Subcommittee of the International Union of Nutritional Sciences since 1986. He also serves as the Section Editor for Nutrition of *Aquaculture*.

Among other awards, Dr. Wilson received the Research Appreciation Award in 1981 and the Distinguished Service Award in 1984 from Catfish Farmers of America. In 1989, he received the first Mississippi Corporation Award of Excellence for outstanding research work in the Mississippi Agriculture and Forestry Experiment Station.

Dr. Wilson's major research area is fish nutrition, with emphasis on the nutrient requirements of the channel catfish. He has presented several invited lectures at national and international meetings and authored or co-authored in excess of 185 publications and presentations, most of which deal with fish nutrition.

CONTRIBUTORS

Shigeru Arai, Ph.D.
Fish Nutrition Division
National Research Institute of Aquaculture
Fisheries Agency
Mie, Japan

Mali Boonyaratpalin, Ph.D.
Department of Fisheries
National Institute of Coastal Aquaculture
Songkhla, Thailand

Thomas Brandt, Ph.D.
National Fish Hatchery and Technology
   Center
U.S. Fish and Wildlife Service
San Marcos, Texas

Newton Castagnolli, Ph.D.
Department of Animal Husbandry
Estadual Paulista University
Jaboticabal, Brazil

C. Young Cho, Ph.D.
Department of Nutritional Sciences
University of Guelph
Guelph, Ontario, Canada

Marie-Francoise Coustans, Ph.D.
DRV Department
IFREMER
Paris, France

Colin Cowey, D.Sc.
Department of Nutritional Sciences
University of Guelph
Guelph, Ontario, Canada

Lin Ding, Ph.D.
Department of Nutritional Sciences
University of Guelph
Guelph, Ontario, Canada

Barbara Grisdale-Helland, Ph.D.
Akvaforsk
Sunndalsora, Norway

Jean Charles Guillaume, Ph.D.
DRV Department
IFREMER
Paris, France

Ronald Hardy, Ph.D.
Utilization Research Division
Northwest Fisheries Science Center
National Marine Fisheries Service
Seattle, Washington

Ståle Helland, Ph.D.
Akvaforsk
Sunndalsora, Norway

Silas Hung, Ph.D.
Department of Animal Science
University of California
Davis, California

Malcolm Jobling, Ph.D.
Norwegian College of Fishery Science
University of Tromsø
Tromsø, Norway

Akio Kanazawa, Ph.D.
Faculty of Fisheries
Kagoshima University
Kagoshima, Japan

George Wm. Kissil, Ph.D.
Nutrition Department
National Center for Mariculture
Israel Oceanographic and Limnological
   Research
Eilat, Israel

Chhorn Lim, Ph.D.
Tropical Aquaculture Research Unit
U.S.D.A.-A.R.S.-P.W.A.
Hawaii Institute of Marine Biology
Kaneohe, Hawaii

Pierre Luquet, Ph.D.
Director of Research
Oceanographic Research Center
Abidjan, Ivory Coast

Robert Métailler, Ph.D.
DRV Department
IFREMER
Paris, France

Jeannine Person-Le Ruyet, Ph.D.
DRV Department
IFREMER
Paris, France

Jean Robin, Ph.D.
DRV Department
IFREMER
Paris, France

Edwin Robinson, Ph.D.
Delta Branch Research and Extension
    Center
Mississippi State University
Stoneville, Mississippi

Shuichi Satoh, Ph.D.
Department of Aquatic Biosciences
Tokyo University of Fisheries
Tokyo, Japan

Sadao Shimeno, Ph.D.
Department of Cultural Fisheries
Faculty of Agriculture
Kochi University
Nankoku, Japan

Trond Storebakken, D.Sc.
Akvaforsk
Sunndalsora, Norway

Robert Wilson, Ph.D.
Department of Biochemistry and Molecular
    Biology
Mississippi State University
Mississippi State, Mississippi

TABLE OF CONTENTS

ARCTIC CHAR, *SALVELINUS ALPINUS* L.

Malcolm Jobling

INTRODUCTION

Arctic char has a northern circumpolar distribution, and anadromous populations occur in Alaska, Canada, Greenland, Iceland, Svalbard, Novaja Zemlja, Siberia, and northern Norway. Nonmigratory and landlocked populations are found both in these northern areas and in areas of North America and Europe further to the south. From the viewpoint of farming, it is the anadromous populations that have been examined with greatest interest, but prior to the upswing of interest in the potential of char as a farmed species, most scientific studies had been carried out on southern nonmigratory populations.

Interest in the farming of char increased during the 1970s and 1980s, primarily with the aim of developing sea-cage rearing of Arctic char in northern temperate regions. Fish from some anadromous populations grow to several kilograms in weight, and early results suggested that growth of char could be rapid during the freshwater rearing phase,[1] which gave rise to optimistic prognoses with respect to the development of commercial production. Rapid growth of Arctic char in fresh water has been subsequently confirmed,[2] but the intensive feeding regimes employed under farming conditions may also promote early maturation of the fish, particularly the males.[3]

Investigations carried out in order to test the ability of char to survive in the sea gave results that were less promising than those obtained under the freshwater rearing phase. Mortalities amongst fish held in sea cages commenced in late summer/early autumn (August-September) and very few fish survived the winter.[4] Results from other studies also suggest a relatively poor seawater survival during the winter months.[5] In the wild, anadromous char are found in the sea, usually close inshore within fjord systems, during the summer months, and the duration of this stay in the marine environment rarely exceeds 50 d. Thus, the increases in mortality observed during autumn in fish held in sea cages are undoubtedly linked to seasonal changes in hypoosmoregulatory ability. Hypoosmoregulatory ability has been found to increase during spring/early summer and then decline during late summer/early autumn, remaining low throughout the winter months.[6]

Problems related to early maturation and seawater survival have limited the development of farming of Arctic char in sea cages, and most commercial producers currently adopt alternative technologies. Farming, on a relatively small scale, is carried out in Canada, Iceland, Sweden, and Norway, and annual production is of the order of a few hundred tonnes, but production is expected to increase within the next few years.

LABORATORY CULTURE

A prerequisite for the optimal design of studies of nutritional requirements of fish is a knowledge about the size of feed particles to be fed, since feed-particle size can affect acceptance, growth, and feed efficiency. The effects of feed-particle size on the growth, and feed efficiency of juvenile Arctic char (3 to 20 g) have been examined.[7] Recommended feed particle sizes correspond to 1.5 to 1.8% of the fork length for small fish (7 to 11 cm), increasing to approximately 2.4% of the fork length for fish of 12 to 13 cm. These feed-particle sizes for char are slightly smaller than those recommended for Atlantic salmon, since growth of small salmon (4 to 20 cm) was optimal when they were fed feed particles corresponding to 2.2 to 2.6% of the fork length.[8]

It should be noted that the char used in Tabachek's experiments[7] had relatively low growth rates and poor feed efficiency.[7] This poor growth performance was also observed in previous studies using the same rearing system.[9,10] The rearing system consisted of tanks with a screen bottom, and thus fish were forced to take food while it fell through the water column. Food intake and growth are lower when char are forced to feed in the water column than when fish are allowed access to food lying on the tank bottom,[11] so the poor growth observed by Tabachek[7,9,10] was probably a direct effect of suboptimal rearing conditions.

NUTRITIONAL STUDIES

Relatively few studies have been carried out in order to elucidate the nutritional requirements of Arctic char, and investigations of char under laboratory conditions have concentrated upon the influences of abiotic (temperature, light) and biotic (stocking density, size-sorting) factors on growth. One reason for the lack of emphasis on nutritional studies is that char have generally shown good rates of growth when fed on commercially available dry diets developed for other salmonid species (Atlantic salmon and rainbow trout), so there has been little pressure to develop dietary formulations specifically for this species. Dietary protein requirements of char appear to be similar to those of other salmonids,[12] and in a study of protein and energy utilization in char, the greatest weight gain and feed utilization was achieved with a diet containing 54% protein and 20% lipid.[10] Lowest feed cost per weight gain was, however, obtained using a diet containing 44% protein and 20% lipid. These studies were conducted using juvenile fish (5 to 100 g) reared in fresh water at temperatures of 10 to 12°C. Tabachek[9] has evaluated the suitability of a series of commercially available grower diets (trout or salmon feeds) for use in char culture. Six different feeds and two strains of char were used in the experiments. Growth and feed efficiency recorded for the fish fed the various diets differed markedly between strains, and it was concluded that the nutritional requirements of the two strains of char were different. The validity of this conclusion is, however, open to question, since rearing conditions were apparently suboptimal. Growth and feed efficiency of the Hammerfest strain of char (the same strain as studied by Jobling[2] and Jørgensen and Jobling[11]) were low, and growth of this strain can be suppressed if fish are forced to feed in the water column. Growth of the other, Labrador, char was faster than that of the Hammerfest strain, and feed efficiency was also better. To what extent the observed differences in growth and feed efficiency indicate differences in nutritional requirements between the char is impossible to assess, since important factors, such as possible differences in preferred modes of feeding behavior (pelagic vs. benthic) and feed availability (related to sinking rates of the various pellet types) are unknown.

The muscle of wild-caught landlocked char contains neutral lipids predominated by 16:0, 16:1, and 18:1 fatty acids.[13] The 16:0 and 18:1 fatty acids, along with 20:1, were also prevalent in the neutral lipids of start-feeding fry, which were the offspring of hatchery-reared parents,[14] even though eggs of both wild and hatchery-reared char are rich in polyunsaturated fatty acids (PUFAs).[15] The PUFA content of egg lipids of char may reach 35% or more, with the relative proportions of 20:5n-3 and 22:6n-3 being quite high. It is clear that the PUFAs are preferentially deposited in the polar lipids, which comprise 40% of the total lipids in the start-feeding fry. Once the fish are weaned onto formulated feeds, the neutral lipid fatty acid profiles change and gradually come to reflect those of the diet.[13] Since many commercially available salmonid feeds contain oils of marine origin, the neutral lipids of hatchery-reared char may contain relatively high levels of 18:1, 20:1, 22:1, and the n-3 series of fatty acids typical of marine food chains.

The natural prey of char in freshwater include crustaceans, insects, and molluscs, with both planktonic and benthic species being represented.[16] Larger char may, in addition, consume fish, and fish such as sandeels, *Ammodytes* spp., and juvenile herring, *Clupea harengus*, may be important components of the diet of char during their summer sojourn in marine coastal waters. Crustaceans, such as krill and amphipods, are also consumed in relatively large numbers,[17] and carotenoid pigments derived from the food are responsible for the characteristic coloring of the tissues in char and other salmonids.

In common with other wild salmonids,[18] the carotenoids found in the tissues of char are either free astaxanthin or esters thereof, with only free astaxanthin being found in the muscle.[19] The muscle of wild-caught char often looks redder than that of other salmonids, and from the limited data available, it appears that the carotenoid content of char muscle is towards the higher end of the salmonid range.[18,19] This red flesh coloration is deemed desirable by the consumer and attempts are, therefore, made to produce farmed char with the deep-red color characteristic of the wild fish. When commercially available dry feed containing synthetic pigment sources have been fed, problems have been experienced in producing fish with this flesh quality. For example, muscle tissue of 2+ char contained only 2 µg pigment per gram following feeding for 9 weeks on a diet containing 40 µg canthaxanthin per gram of feed.[20] This pigment concentration is far below the 3 to 4 µg/g considered an acceptable concentration in the flesh of marketed salmonids.[18] The digestibility (18% for 1+, and 39% for 2+ fish) of synthetic pigment (canthaxanthin) reported for char[20] appears to be lower than that found in studies with other salmonids,[18] so this may partially explain why problems have arisen. There are, however, large variations in the results of the digestibility trials, and methodological problems associated with the extraction and analysis of carotenoids makes direct comparisons between studies difficult. Attempts have been made to compensate for the poor digestibility and retention of pigment by increasing the pigment concentration in commercial dry feeds, and feeds used in char farming may be formulated to contain 80 to 100 µg astaxanthin per gram.

REFERENCES

1. **Gjedrem, T. and Gunnes, K.,** Comparison of growth rate in Atlantic salmon, pink salmon, Arctic charr, sea trout and rainbow trout under Norwegian farming conditions, *Aquaculture*, 13, 135, 1978.
2. **Jobling, M.,** Influence of body weight and temperature on growth rates of Arctic charr, *Salvelinus alpinus* (L.), *J. Fish Biol.*, 22, 471, 1983.
3. **Nordeng, H.,** Solution to the 'char problem' based on Arctic char (*Salvelinus alpinus*) in Norway, *Can. J. Fish. Aquat. Sci.*, 40, 1372, 1983.
4. **Gjedrem, T.,** Survival of Arctic charr in the sea during fall and winter, *Aquaculture*, 6, 189, 1975.
5. **Wandsvik, A. and Jobling, M.,** Overwintering mortality of migratory Arctic charr, *Salvelinus alpinus* (L.), reared in salt water, *J. Fish Biol.*, 20, 701, 1982.
6. **Finstad, B., Nilssen, K. J., and Arnesen, A. M.,** Seasonal changes in sea-water tolerance of Arctic charr (*Salvelinus alpinus*), *J. Comp. Physiol.*, B159, 371, 1989.
7. **Tabachek, J. L.,** The effect of feed particle size on the growth and feed efficiency of Arctic charr [*Salvelinus alpinus* (L.)], *Aquaculture*, 71, 319, 1988.
8. **Wankowski, J. W. J. and Thorpe, J. E.,** The role of food particle size in the growth of juvenile Atlantic salmon (*Salmo salar* L.), *J. Fish Biol.*, 14, 351, 1979.
9. **Tabachek, J. L.,** Evaluation of grower diets for intensive culture of two strains of Arctic charr (*Salvelinus alpinus* L.), *Can. Tech. Rep. Fish. Aquat. Sci.*, 1281, 21, 1984.
10. **Tabachek, J. L.,** Influence of dietary protein and lipid levels on growth, body composition and utilization efficiencies of Arctic charr, *Salvelinus alpinus* L., *J. Fish Biol.*, 29, 139, 1986.
11. **Jørgensen, E. H. and Jobling, M.,** Feeding modes in Arctic charr, *Salvelinus alpinus* L.: the importance of bottom feeding for the maintenance of growth, *Aquaculture*, 86, 379, 1990.
12. **Jobling, M. and Wandsvik, A.,** Quantitative protein requirements of Arctic charr, *Salvelinus alpinus* (L.), *J. Fish Biol.*, 22, 705, 1983.

13. **Ringø, E. and Nilsen, B.,** Hatchery-reared landlocked Arctic charr, *Salvelinus alpinus* (L.), from Lake Takvatn, reared in fresh and sea water. I. Biochemical composition of food, and lipid composition of fish reared in fresh water, *Aquaculture*, 67, 343, 1987.

14. **Christiansen, J. S., Ringø, E., and Jobling, M.,** Effects of sustained exercise on growth and body composition of first-feeding fry of Arctic charr, *Salvelinus alpinus* (L.), *Aquaculture*, 79, 329, 1989.

15. **Ringø, E.,** personal communication, 1989.

16. **Johnson, L.,** The Arctic charr, *Salvelinus alpinus*, in *Charrs, Salmonid Fishes of the Genus Salvelinus*, Balon, E. K., Ed., Junk, The Hague, 1980, 15.

17. **Grønvik, S. and Klemetsen, A.,** Marine food and diet overlap of co-occurring Arctic charr, *Salvelinus alpinus* (L.), brown trout, *Salmo trutta* L., and Atlantic salmon, *S. salar* L., off Senja, N. Norway, *Polar Biol.*, 7, 173, 1987.

18. **Torrissen, O. J., Hardy, R. W., and Shearer, K. D.,** Pigmentation of salmonids — Carotenoid deposition and metabolism, *Rev. Aquat. Sci.*, 1, 209, 1989.

19. **Scalia, S., Isaksen, M., and Francis, G. W.,** Carotenoids of the Arctic charr, *Salvelinus alpinus* (L.), *J. Fish Biol.*, 34, 969, 1989.

20. **Christiansen, J. S. and Wallace, J. C.,** Deposition of canthaxanthin and muscle lipid in two size groups of Arctic charr, *Salvelinus alpinus* (L.), *Aquaculture*, 69, 69, 1988.

ASIAN SEABASS, *LATES CALCARIFER*

Mali Boonyaratpalin

INTRODUCTION

Asian seabass, also called giant sea perch, is an economically important food fish and sport fish in the tropical and subtropical areas of the western Pacific and Indian Ocean countries, including India, Burma, Sri Lanka, Bangladesh, Malay Peninsula, Java, Borneo, Celebes, Philippines, Papua New Guinea, Northern Australia, Southern China, and Taiwan.

Natural production of seabass fry has been declining due to overfishing and destruction of habitat by urban, rural, agricultural, and industrial development. Seabass are a fast-growing fish, attaining marketable size within 8 months, generally with a growth rate of l kg/year, with white flesh that is quite acceptable for eating by most people, and brings a high market price. Seabass also have many characteristics that are favorable for coastal aquaculture, i.e., they can grow well in water of high turbidity and varying salinities, and can tolerate rough handling and the crowded conditions of net-cage culture. They are easily tamed for aquaculture and accept feeding by humans. Therefore, seed production and culture programs were established in many countries.

Seabass had been cultured in Hong Kong, Indonesia, Malaysia, Philippines, Singapore, Taiwan, and Thailand for many years. Data published by SEAFDEC[1] show that seabass production was 34, 1384, 1067, 219, and 1158 t in Hong Kong, Indonesia, Malaysia, Singapore, and Thailand, respectively in 1987.

Three methods of spawning are used in Thailand for the mass production of seabass fry. Artificial fertilization involves catching spawners in estuaries and stripping the males and females for eggs and milt. The fertilized eggs are then transported to the hatchery for subsequent incubation. Induced spawning by hormone injection is practiced where natural spawning in the tank does not take place or when an early spawning is needed. Natural spawning in tanks is the easiest and best method for commercial-scale seabass seed production. This is because it produces the largest quantity of eggs, of the highest quality, and for a prolonged period.

Larval rearing is one of the most important steps after seed production. Many factors must be considered at this stage to avoid high mortality and to maintain high production. Salinity should be between 25 and 30 ‰ for larvae, fry, and food organism culture. During the first 15 d, salinity is maintained at about 20 to 28 ‰. Fry can tolerate freshwater starting when they reach 4.5 mm total length, but growth is reduced. On reaching juvenile stage (13.9 to 17.7 mm), fish begin to grow as well in fresh water as in seawater. Food and feeding are the most important factors affecting growth, health, size variation, and survival of seabass larvae and fry. Many kinds of food are fed to the larvae and fry, depending on their age or size. Rotifers are the most suitable food for larvae at the early stages of rearing. Microcrustacea (*Artemia* and *Moina*) are usually fed once fry reach 4 mm in size.

The growout phase involves the rearing of seabass from fingerling (10 cm) to marketable size. The marketable size of seabass in central Thailand is 500 to 800 g. This size brings a higher price than smaller or larger fish. However, marketable size in southern Thailand is 600 to 800 g or 1200 to 2000 g. The culture period varies from 5 to 8 months (for 500 to 800 g fish) to as much as 12 to 20 months for larger fish. Seabass are culture in ponds and net cages. Production yields range from 0.87 to 0.91 kg/m^3/crop in ponds to as high as 35 to 40 kg/m^3/crop in net cages.

NUTRIENT REQUIREMENTS

The traditional feed for most of the commercial seabass growout culture operations in Southeast Asia is raw, minced, or chopped trash fish. This traditional feed often does not completely satisfy the nutritional requirements of seabass, resulting in many malnutrition problems and low survival. Furthermore, the supply of trash fish is becoming limited due to increased demand from other aquaculture activities. Market price, quality, and quantity of trash fish is often uncertain due to changes in the climate, season, or conditions of handling and storage. For these reasons, efforts have been made by researchers in Indonesia,[2] Singapore,[3-5] Tahiti,[6] and Thailand[7-13] to study nutrient requirements and to develop complete artificial feeds that will supply seabass with essential nutrients and improve profits for culturists.

The nutritional requirements of seabass are similar to other marine carnivorous fish in respect to quality of protein, amino acids, lipids, fatty acids, carbohydrates, vitamins, and minerals. The quantitative demand for these nutrients varies depending on species, growth stage, and environmental conditions.

PROTEIN AND AMINO ACIDS

The dietary protein requirement of seabass was initially estimated to be 45 to 55%.[6] This study involved feeding four practical diets containing 35 to 55% crude protein mainly from Norwegian fish meal and a control diet containing 52% crude protein consisting of fish meal and fish protein concentrate. The protein to energy ratio was held constant at about 140 mg protein/kcal. The highest growth rate was obtained with the control diet. Sakaras et al.[11] evaluated six practical diets with three dietary protein levels of 45, 50, and 55% and two lipid levels of 10 and 15% at each protein level in 7.47 g seabass. Fish were fed to satiation twice daily for 8 weeks. The best growth rate, feed conversion, protein retention, and protein efficiency ratio was observed in fish fed the diet containing 50% crude protein, 15% lipid, and 7.33 kcal/g of protein. In a subsequent study, fish fed a diet containing 45% crude protein and 18% lipid had the highest growth rate.[12] The optimal dietary protein is usually higher for larval and fry stages and lower for the growout stage. The optimal dietary protein level for growout seabass has been reported to be 40 to 45% crude protein in diets containing 12% lipid.[14]

Animal proteins such as fish meal and squid meal are the best protein sources and the most readily accepted by seabass. The replacement of 5% fish meal with an equal amount of squid meal in a seabass fingerling diet improved growth and feed efficiency by 11.44 and 13.3%, respectively.[15]

No information is available on the essential amino acid requirements of seabass. However, it has been shown that excessive dietary tyrosine may result in kidney disease.[16]

LIPIDS AND ESSENTIAL FATTY ACIDS

No differences in growth rate and survival were observed when seabass were fed three diets containing 52% crude protein and 6, 10, or 14% lipid.[6] The protein sparing effect of lipid was observed in a study that found that the optimum dietary lipid level of seabass fingerlings was 15 and 18% at protein levels of 50 and 45%, respectively.[11,12]

Signs of an essential fatty acid (EFA) deficiency were observed in seabass fingerlings after being fed a diet containing 0.46% n-3 highly unsaturated fatty acids (HUFA) for 2 weeks. The initial deficiency sign was a reddening of the fins and skin, which was followed by other deficiency signs, such as abnormal eyes, a shock syndrome, loss of appetite, poor growth, and swollen, pale livers. Fish fed a diet containing 0.88% n-3 HUFA also showed slight EFA deficiency, such as reddening of the fins and skin. Fish fed a diet containing 1.72% n-3 HUFA, approximately 13% of the dietary lipid, had the best growth rate and showed no signs of EFA deficiency.[10] In a subsequent study, six fish meal and casein diets containing levels of n-3 HUFA ranging from 1 to 2% of the diet were fed for 10 weeks. Dietary n-3 HUFA levels were adjusted using squid liver oil. No difference in growth, feed efficiency, and mortality was detected and no signs of EFA deficiency were observed.[17] The results of the two studies indicate that the dietary n-3 HUFA requirement for seabass fingerlings is 1.0 to 1.7% of the diet.

VITAMINS

Not all vitamins appear to be essential in practical seabass feeds, even though they are required for normal metabolic functions. Some vitamins may be synthesized by seabass in almost sufficient quantities to meet requirements or may be present in adequate amounts in the practical feed ingredients. Vitamin requirements depend upon size of fish, stage of sexual maturity, growth rate, environmental conditions, and dietary nutrient interrelations. The vitamin requirements of seabass seem to decrease as fish size increases.[18]

The addition of a vitamin mix to trash fish fed to seabass fingerlings increased growth rate from 9.36 to 23.48 g and reduced the feed conversion ratio from 7.44 to 3.83 during a 9-week rearing period.[19] The vitamin requirements for seabass have been established for young fish by supplementing both practical and semipurified diets with various levels of a specific vitamin. The requirements have been based on the minimum dietary vitamin level that will support maximum growth or maximum tissue storage and prevent deficiency signs. No differences in weight gain, feed efficiency, and total mortality were observed when seabass were fed practical diets without the addition of choline, niacin, inositol, or vitamin E.[20] Similarly, no differences in growth rate, feed efficiency, and survival were observed in seabass fingerlings fed practical diets without supplemental pyridoxine and pantothenic acid.[21] Lower weight gain and feed efficiency were observed in fish fed diets without supplemental thiamin and riboflavin after 60 d. Fish fed diets without added vitamin C had normal growth rates for about 15 d, after which growth ceased and deficiency signs, such as poor appetite, loss of equilibrium, gill hemorrhage, and scoliosis, developed. Severe mortality occurred after 45 d and all vitamin C-deficient fish died within 60 d.

Two studies have been conducted to determine the quantitative vitamin C requirement of seabass fingerlings.[9,18] The first study involved feeding practical diets supplemented with 0, 500, 1000, 1500, 2000, and 2500 mg vitamin C/kg of diet to seabass in fresh water for 10 weeks. The best growth rates were observed in fish fed diets containing 1000 mg vitamin C/kg or higher. In the second study, fish were fed practical diets containing 0, 500, 700, 900, 1100, 2500, and 5000 mg vitamin C/kg diet in seawater for 8 weeks. Fish fed diets with vitamin C levels of 500+ mg/kg of diet had satisfactory growth rates, with only slight vitamin C deficiency signs being observed in fish fed the 500 mg vitamin C/kg diet. The vitamin C content of liver and kidney increased with the increase in dietary vitamin C levels. In conclusion, the minimum level of supplemental vitamin C required for normal growth of seabass fingerlings in seawater was 500 mg/kg diet, and a supplemental level of 1100 mg/kg diet was required for normal tissue storage.

An example of a semipurified diet that has been used in vitamin studies for seabass fingerlings is presented in Table 1. When this diet was used to determine the pyridoxine requirement, the results indicated that 5 mg pyridoxine/kg diet was required for normal growth and 10 mg pyridoxine/kg diet was needed for normal lymphocyte levels.[13] This semipurified diet has also been used to determine the essentiality of certain vitamins and to document their deficiency signs in seabass as summarized in Table 2.[22]

CARBOHYDRATES

The natural foods of seabass are high in protein, so it is assumed that they do not utilize carbohydrates well. We have observed in our laboratory that the addition of small amounts of starch (10%) to diets lacking carbohydrate improved growth rates in seabass, but higher levels (27%) reduced growth rates. Thus, until additional research has been conducted, it seems

TABLE 1
Example of a Semipurified Diet for Seabass[13]

Ingredient	Percent
Casein (viamin free)	50
Gelatin	10
Cod liver oil	6
Soybean oil	3
Gelatinized starch	5
Cellulose	8
Carboxymethyl cellulose	5
Vitamin mixture[a]	2
McCollum's salt mixture[b]	4
Amino acid mixture[c]	7
Water (ml)	80

[a] To contain as mg/100 g dry diet: thiamine HCl, 5; riboflavin, 20; choline chloride, 500; nicotinic acid, 75; Ca-pantothenate, 50; pyridoxine HCl, 5; inositol, 200; biotin, 0.5; folic acid, 1.5; vitamin B_{12}, 0.1; menadione, 4.0; α-tocopherol acetate, 40; vitamin A, 1000 IU; vitamin D_3, 200 IU; BHT, 1.0; ascorbic acid, 100.

[b] McCollum's salt mixture no. 185 plus trace elements (units/100 g mineral mixture): calcium lactate, 32.70 g; K_2HPO_4, 23.98 g; $CaHPO_4 \cdot 2H_2O$, 13.58 g; $MgSO_4 \cdot 7H_2O$, 13.20 g; $Na_2HPO_4 \cdot 2H_2O$, 8.72 g; NaCl, 4.35 g; ferric citrate, 2.97 g; $ZnSO_4 \cdot 7H_2O$, 0.3 g; $CoCl_2 \cdot 6H_2O$, 100 mg; $MnSO_4 \cdot H_2O$, 80 mg; KI, 15 mg; $AlCl_3 \cdot 6H_2O$, 15 mg; $CuCl_2$, 10 mg.

[c] Amino acid mixture[24] (g/100 g dry diet): L-phenylalanine, 0.6; L-arginine·HCl, 1.3; L-cystine, 0.7; L-tryptophan, 0.2; L-histidine·HCl·H_2O, 0.2; DL-alanine, 1.3; L-aspartic acid·Na, 1.0; L-valine, 0.7; L-lysine·HCl, 0.6; glycine, 0.4.

reasonable to suggest that the total carbohydrate content of feeds for seabass fingerlings not exceed 20%.

MINERALS

Mineral requirements for seabass have not been adequately evaluated because seabass do not readily accept purified diets. Experiments on the optimum level of supplemental mineral mix UPS XII and the effect of microminerals and calcium lactate on the growth of seabass fingerlings were done by feeding five fish meal and casein-based diets containing either 0, 2, or 4% UPS XII mineral mix, 4% macrominerals, or 4% calcium lactate free macrominerals.[24] The results indicated that the diet containing 2% UPS XII mineral mix gave optimum growth and that seabass appear to obtain adequate microminerals from their water supply or practical feedstuffs. The available phosphorus requirement for seabass was found to be 0.65%.[25]

FEED FORMULATION AND FEEDING PRACTICES

Seabass are willing to accept food from the water column, but not at the surface or from the bottom, and prefer a soft, moist feed. Therefore, before being fed to seabass, extruded floating feeds are normally soaked in fresh water until they become saturated, slow sinking pellets. The successful use of dry pelleted feeds for the culture of seabass in cages has been reported.[26] However, moist pellets are inexpensive and easy for farmers to process, but a

TABLE 2
Vitamin Requirements and Deficiency Signs[9,13,18,22]

Vitamin	Requirement mg/kg diet	Deficiency signs
Thiamine	R[a]	Poor growth, substantial post-handling shock, high mortality
Riboflavin	R	Erratic swimming, cataracts
Pyridoxine	5	Avoidance of schooling, erratic spiral swimming, surfacing, lesion of lower lip, high mortality, convulsions
Pantothenic acid	R	Ventral fin hemorrhage and erosion, clubbed gill, high mortality
Nicotinic acid	NA[b]	
Biotin	NA	
Inositol	R	Poor growth, abnormal bone formation
Choline	NA	
Folic acid	NA	
Ascorbic acid	700	Bleeding gill, short operculum, short snout, exopthalmia, short body, loss of equilibrium
Vitamin A	NA	
Vitamin D	NA	
Vitamin E	R	Muscular atrophy, susceptible to disease
Vitamin K	NA	

[a] Required but quantitative requirement not established.
[b] No information available.

binder is required. The most commonly used binder is starch or cooked starch. Though extruded feeds are easy to use, they are more expensive than other types due to high manufacturing costs and because an excess of heat-sensitive vitamins must be added to the feed. The feed conversion ratio of extruded feed or moist pellets is 1.2 to 1.0 (dry weight basis).

The traditional feed or trash fish supplemented with vitamins and minerals is recommended to farmers who can obtain fresh trash fish easily and at a low cost. The feed conversion ratio of trash fish feed is about 4:1.

Based on the results of research carried out by the Department of Fisheries in Thailand, an optimal feed formulation has been recommended to seabass farmers (Table 3). It has been well accepted and gives favorable growth results. Seabass are typically fed based on a percentage of the standing crop weight. Feeding rates are affected by fish size and, typically, smaller fish will eat more feed in relation to their size than larger fish. Table 4 provides suggested feeding rates for small seabass. These feeding rates are only guides. The fish culturist must use considerable judgment, since many factors are involved. Satiation feeding is recommended for seabass culture because of the cannibalistic and aggressive characteristics of the fish. Feeding the fish slowly to satiation will result in more even growth in the fish population, minimizing mortality due to cannibalism and the need for frequent grading.

Medicated feeds can be of any type and any formulation with the appropriate chemical or drug incorporated into the formula prior to processing or pelleting. In Thailand, extruded feeds are commonly top-dressed with medications immediately after extrusion or soaked in medicated water just prior to feeding.

TABLE 3
Example of Practical Diet for Seabass[27]

Ingredient	Percent
Fish meal	70.00
Rice bran	12.40
Vitamins	1.00
Minerals	2.00
Gelatinized starch	10.00
Ascorbic acid	0.10
Marine fish oil	1.50
Vegetable oil	3.00
Water	40–50

TABLE 4
Suggested Feeding Rates for Small Seabass[27]

Fish size (g)	Feeding frequency (times/d)	Feeding rate (% body weight/d)
1.8–5.4	2–3	7.18
5.5–11.5	2–3	5.70
11.6–19.2	2	4.59
19.3–27.9	2	3.90
28.0–45.0	2	3.50

REFERENCES

1. **SEAFDEC,** Fisheries Statistical Bulletin for South China Seas and Asia, Southeast Asian Fisheries Development Center, Bangkok, Thailand, 1987, 100.
2. **Tacon, F. G. J. and Rausin, N.,** Seabass cage culture trial, in INS/81/008/Technical Paper 13, The Food and Feeding of Seabass *Lates calcarifer,* Grouper *Epinephelus taurina* and Rabbitfish *Siganus canaliculatus* in Floating Net Cages, The National Seafarming Development Center, Lampung, Indonesia, 1989, 34.
3. **Chou, R.,** The effect of dietary water content on the feed intake, food conversion efficiency and growth of young seabass (*Lates calcarifer*), *Singapore J. Primary Ind.,* 12, 120, 1984.
4. **Chou, R.,** Progress of seafarming activities, research and development in Singapore, Appendix XIII, in *Report of the Second National Coordinators Meeting on Regional Seafarming Development and Demonstration Project* RAS/86/024, 20–23 September, Singapore Regional Seafarming Development and Demonstration Project, Network of Aquaculture Centers in Asia, Bangkok, Thailand, 1988, 81.
5. **Chou, R., Wong, F. J., and Lee, H. B.,** Observations on the weaning of fingerlings of some cultured marine food fish in Singapore, *Singapore J. Primary Ind.,* 15, 64, 1987.
6. **Cuzon, G.,** Preliminary nutritional studies of seabass *Lates calcarifer* (Bloch) protein and lipid requirements, in 19th Annual Conference and Exposition, World Aquaculture Society, Hawaii '88 Program and Abstracts, 1988, 15.
7. **Boonyaratpalin, M.,** *Annual Report of Fish Nutrition Project in Thailand,* National Inland Fisheries Institute, Department of Fisheries, Bangkok, Thailand, 1988.
8. **Boonyaratpalin, M., Pechamanee, T., Chungyamplin, S., and Buranapanidgit, J.,** Seabass feed, feeding and nutrition deficiency signs, Extension Paper No. 1, National Institute of Coastal Aquaculture, Department of Fisheries, Songkhla, Thailand, 1989 (in Thai).
9. **Boonyaratpalin, M., Unprasert, N., and Buranapanidgit, J.,** Optimal supplementary vitamin C level in seabass fingerling diet, in *The Current Status of Fish Nutrition in Aquaculture,* Takeda, M. and Watanabe, T., Eds., Tokyo University of Fisheries, Tokyo, Japan, 1989, 149.
10. **Buranapanidgit, J., Boonyaratpalin, M., Watanabe, T., Pechmanee, T., and Yashiro, R.,** Essential fatty acid requirement of juvenile seabass *Lates calcarifer,* Technical Paper No. 3, National Institute of Coastal Aquaculture, Department of Fisheries, Thailand, 1988 (in Thai).

11. **Sakaras, W., Boonyaratpalin, M., Unprasert, N., and Kumpang, P.,** Optimum dietary protein energy ratio in seabass feed I, Technical Paper No. 7, Rayong Brackishwater Fisheries Station, Thailand, 1988 (in Thai).

12. **Sakaras, W., Boonyaratpalin, M., and Unprasert, N.,** Optimum dietary protein energy ratio in seabass feed II, Technical Paper No. 8, Rayong Brackishwater Fisheries Station, Thailand, 1989 (in Thai).

13. **Wanakowat, J., Boonyaratpalin, M., Pimoljinda, T., and Assavaaree, M.,** Vitamin B6 requirement of juvenile seabass *Lates calcarifer*, in *The Current Status of Fish Nutrition in Aquaculture*, Takeda, M. and Watababe, T., Eds., Tokyo University of Fisheries, Tokyo, Japan, 1989, 141.

14. **Wong, F. J. and Chou, R.,** Dietary protein requirement of early grow-out seabass (*Lates calcarifer* Bloch) and some observations on the performance of two practical formulated feeds, *Singapore J. Primary Ind.*, 17, 134, 1989.

15. **Chungyampin, S. and Boonyaratpalin, M.,** Effect of type of feeds and squid meal on growth of seabass fingerlings, National Institute of Coastal Aquaculture, Songkhla, Thailand, 1988, unpublished data.

16. **Boonyaratpalin, S., Fryer, J. L., Hedrick, R. P., Supamattaya, K., Direkbusarakom, S., and Jadesadakraisorn, U.,** Development of kidney disease and occurrence of pathogens in seabass (*Lates calcarifer*) fed two different diets, Technical Paper No. 8, National Institute of Coastal Aquaculture, Department of Fisheries, Thailand, 1990.

17. **Buranapanidgid, J., Boonyaratpalin, M., and Kaewninglard, S.,** Optimum level of ω3HUFA on juvenile seabass, *Lates calcalifer*, in *IDRC Fish Nutrition Project Annual Report*, Department of Fisheries, Thailand, 1989.

18. **Boonyaratpalin, M., Unprasert, N., Kosutharak, P., Sothana, W., and Chumsungnern, S.,** Effect of different levels of vitamin C added in diet on growth, feed efficiency and survival rate of seabass, Technical Paper No. 6. National Institute of Coastal Aquaculture, Department of Fisheries, Thailand, 1989 (in Thai).

19. **Phromkhuntong, W., Supamattaya, K., and Jittione, W.,** Effect of water-soluble vitamins on growth, body composition and histology of seabass, Report of the Aquatic Science Division, Faculty of Natural Resources, Prince of Songkhla University, Thailand, 1987.

20. **Boonyaratpalin, M., Unprasert, N., Kosutharak, P., Chumsungnern, S., and Sothana, W.,** Effect of choline, niacin, inositol and vitamin E on the growth, feed efficiency and survival of seabass fingerling in freshwater, Technical Paper No.7, National Institute of Coastal Aquaculture, Department of Fisheries, Thailand, 1988 (in Thai).

21. **Pimoljinda, T. and Boonyaratpalin, M.,** Study on vitamin requirements of seabass *Lates calcarifer* Bloch in sea water, Technical Paper No. 3, Phuket Brackishwater Fisheries Station, Department of Fisheries, Thailand, 1989 (in Thai).

22. **Boonyaratpalin, M. and Wanakowat, J.,** Effect of thiamine, riboflavin, pantothenic acid and inositol on growth, feed efficiency and mortality of juvenile seabass, National Institute of Coastal Aquaculture, Thailand, 1990, unpublished data.

23. **Yone, Y.,** Nutritional studies of red sea bream, in *Proc. First Int. Conf. Aquaculture Nutr.*, Price, K. S., Shaw, W. N., and Danberg, D. S., Eds., University of Delaware, Lewes/Rehoboth, 1976, 39.

24. **Porn-ngam, N., Chonchuenchob, P., and Boonchum, U.,** Dietary mineral requirement in seabass, in *IDRC Fish Nutrition Project Annual Report*, Department of Fisheries, Thailand, 1989, 3.

25. **Boonyaratpalin, M. and Phongmaneerat, J.,** Requirement of seabass for dietary phosphorus, Technical Paper No. 4, National Institute of Coastal Aquaculture, Thailand, 1990 (in press).

26. **Sakaras, W.,** Experiment on seabass, *Lates calcarifer* (Bloch), cultured in cages with dry pellet, Technical Paper No. 5, Rayong Brackishwater Fisheries Station, Thailand, 1990 (in press).

27. **Boonyaratpalin, M.,** Seabass feed, in *Manual for Training Provincial Fisheries Officers*, Prajuabkhirikhan Coastal Aquaculture Research and Development Center, Thailand, 1988 (in Thai).

ATLANTIC SALMON, *SALMO SALAR*

Ståle Helland, Trond Storebakken, and Barbara Grisdale-Helland

INTRODUCTION

The production of farm-raised Atlantic salmon is a relatively new industry, which was rather limited until the early 1970s. The farming of this species has since escalated to a world production in 1989 of over 177,000 tons,[1] of which Norway produced 65%. Other major producing countries are United Kingdom (18%), Faeroe Islands (5%), Ireland (4%), Chile (3%), Iceland (2%), United States (2%), and Canada (2%). The world production of farm-raised Atlantic salmon is approximately 4.5 times that of Pacific salmon.[1]

The production of Atlantic salmon is divided into a juvenile freshwater stage and a growing/finishing period in saltwater. The freshwater stage lasts for approximately 1.5 years, at the end of which time the fish has smoltified and is ready to move from a hypotonic (fresh water) to a hypertonic environment (seawater).

Smoltification is influenced by environmental factors, such as light and water temperature,[2] and in addition, nutritional factors. Besides the obvious need for an adequate diet conducive to growth that allows the salmon to reach the size needed for smoltification,[3] this process may also be affected by salt level in the feed.[4] These options for manipulation of smoltification can be used to produce salmon that can tolerate the osmotic pressure of saltwater at an age of approximately 6 months. Smoltification is accompanied by significant metabolic changes,[5] indicating that nutrient requirements should be assessed separately for fish in the freshwater period and in the growing/finishing period in the sea.

The duration of the growth period in the sea depends on the desired market size of the fish. Most Atlantic salmon production is targeted at markets that desire fish weighing from 2 to 6 kg. This requires that the salmon are grown in the sea for up to 2 years, at which time the fish will normally start to become sexually mature. This process is associated with a reduction in appetite and a repartitioning of nutrients from other tissues to the gonads. Mature fish are characterized by brownish skin, reduced pigmentation in the flesh, and lower concentrations of protein and fat in the fillet, resulting in a marked decrease in the quality of the product.[6] After 2 years in the sea, the broodstock is traditionally moved back to fresh water for spawning, which normally occurs in late fall and early winter.

Ideally, separate nutrient requirements should be established for fry and fingerlings, smolt, growers, and broodstock. Furthermore, improvements in growth as a result of genetic selection of this species (3 to 4% per year[7]) necessitates continual reevaluation of nutrient requirements.

LABORATORY CULTURE

The domestication of Atlantic salmon began less than 20 years ago. This short period of domestication is reflected in the ease by which these fish are stressed, both under farming conditions and in the laboratory. Fevolden et al.[8] examined the response to stress by measuring blood cortisol levels and found that a given stress caused twice the level of response in Atlantic salmon as that in rainbow trout. A consequence of this ease of excitation is that special care must be taken to reduce environmental stress on the fish. Observations at our institute indicate that growth can be improved by restricting traffic around the tanks, and by elevating the tanks and directing the light source into the tanks so that workers are not seen by the fish. The use of deep, cylindrical tanks has also been beneficial in comparison with the traditional shallow, square tanks, which have a larger surface area.

TABLE 1
Examples of Purified Diets for Atlantic Salmon[14]

Ingredient	Diet 1	Diet 2
Casein	40.0	—
Isolated fish protein	—	40.0
Dextrin, tapioca	20.0	20.0
L-glutamic acid	5.2	5.2
Cellulose	21.3	21.3
Herring oil	10.0	10.0
Carboxymethyl cellulose	1.0	1.0
Vitamin and mineral mix[a]	2.5	2.5

[a]   Vitamins (mg/kg diet): thiamine·HCl, 50; riboflavin, 200; pyridoxine·HCl, 50; niacin, 750; D-Ca-pantothenate, 500; inositol, 2000; D-biotin, 5; folic acid, 15; L-ascorbic acid, 1000; cyanocobalamine, 0.1; menadione sodium bisulfite, 40; choline Cl, 5000; DL-α-tocopherol acetate, 500 IU; vitamin A palmitate, 10,000 IU; vitamin D_3, 4000 IU.
Minerals (mg/kg diet): $CaCO_3$, 250; KCl, 4672; KH_2PO_4, 4000; Na_2HPO_4, 3087; $MgSO_4$, 2475; $FeSO_4·7H_2O$, 250; $ZnSO_4·7H_2O$, 220; $MnSO_4·H_2O$, 92; $CuSO_4$, 20; KI, 1.3; $(NH_4)_6Mo_7O_{24}·4H_2O$, 0.4; $CoCl_2·6H_2O$, 0.2; Na_2SeO_3, 0.2; $CrCl_3·6H_2O$, 0.5.

Light is an important environmental factor to consider in both the laboratory and in practical aquaculture of Atlantic salmon. In the freshwater stage, growth is generally stimulated under conditions of long days, but may result in delayed or incomplete smoltification.[9-13]

In comparison with rainbow trout, Atlantic salmon have a small stomach and, therefore, require more frequent feeding. Our results indicate that fry and fingerlings should receive food every 5 to 10 min,[30] whereas parr may be fed every 20 to 30 min. Larger fish require less frequent feeding.

The establishment of the requirements for many of the nutrients requires the use of purified diets, such as those used by Rumsey and Ketola,[14] to study the effect of amino acid supplementation on the nutritional value of casein in diets fed to Atlantic salmon fry (Table 1). A major problem with such diets has been, however, to obtain growth comparable with that obtained with practical diets. Atlantic salmon frequently will not grow well on this kind of diet, which may be related to problems with appetite.

NUTRIENT REQUIREMENTS

The literature regarding the nutrient requirements of Atlantic salmon is surprisingly scarce, when the economic importance of the farming of this species is taken into account. Almost no nutrient requirements have been established with the certainty that is common for our other farm animals. In this chapter, therefore, we have been forced to include information from sources that have not been subjected to peer review, e.g., abstracts. We have, however, tried to contact authors for verification of the information in the nonreviewed publications.

ENERGY

The energy requirement for maximum growth depends mainly on the size of the fish and the water temperature.[15] The energy requirement for maximum growth under practical conditions is around 390 kJ digestible energy (DE)/kg fish/d for 10 g fingerlings at 10°C.[16-19] Under these conditions, fingerlings require approximately 17 MJ DE/kg body weight increase.[17] The corresponding value for 1 kg salmon in 10°C saltwater is approximately 150 kJ DE/kg fish/d or 19 to 20 MJ DE/kg body weight increase.[16] (These values may be slightly

TABLE 2
Essential Amino Acid Composition of Whole Body and Eggs of Atlantic Salmon Expressed as Percent of Crude Protein

Amino acid	Whole salmon[20]	Carcass[21]	Immature ovary[22]	Eggs[23]
Arginine	6.6	5.9	12.8	6.1
Cystine	1.0	—	1.2	0.6
Cystine + methionine	—	3.9	—	—
Histidine	3.0	3.3	4.5	2.7
Isoleucine	4.4	5.1	4.3	5.9
Leucine	7.7	7.8	6.9	8.7
Lysine	9.3	8.6	10.5	7.7
Methionine	1.8	3.3	1.6	2.6
Phenylalanine	4.4	4.5	2.8	4.6
Threonine	5.0	3.9	4.3	4.6
Tryptophan	0.9	—	1.0	—
Valine	5.1	5.6	5.9	6.8

higher at the present time because of the genetic improvements in growth that have occurred since the experiments that these estimates are based upon were done.) The calculations are based on a commercial diet with 44% protein, 23% fat, and 16% extruded starch. The maintenance energy requirement of Atlantic salmon has not been established.

PROTEIN AND AMINO ACIDS

The amino acid requirements of Atlantic salmon have not been determined. The essential amino acid profiles of whole fish and eggs, presented in Table 2, however, are similar to those of other salmonids examined.[20] This gives some support for basing recommendations on chemical scores obtained from the composition data and requirements obtained with other species of salmonids.[20,24] As an example of the usefulness of this preliminary approach, Ketola[21] has demonstrated significant growth improvement of Atlantic salmon fry by supplementing soy- or casein-based diets with amino acids according to the composition of whole fish and eggs.

The optimal content of dietary protein for juvenile salmon is approximately 21 g crude protein (CP)/MJ gross energy (GE), whereas growing/finishing fish grow well on high-energy diets containing 17 to 21 g CP/MJ GE (Table 3). Based on essentially the same experiments as listed in Table 3, Austreng et al.[30] suggested that 45 to 50% of the metabolizable energy in a starter diet should be derived from protein. Corresponding values for grower diets and broodstock diets are 40 to 45% and 45 to 50%, respectively. When feeding is optimal and diet composition is according to the above recommendations, whole-body protein retention will be approximately 45% in fingerlings.[17]

LIPIDS

The fatty acid requirement and the requirements of other specific lipid components for Atlantic salmon are not established. Generally, the fatty acids of Atlantic salmon are rich in n-3 long-chain polyenic fatty acids.[31,32] This suggests that the requirement for n-3 fatty acids is high for this species, in agreement with that determined for other salmonids.[24] Estimating the fatty acid requirement directly from the body composition, however, is not recommended. The carcass fatty acid composition is strongly affected by the makeup of the dietary lipids[31-33] and, in addition, may be affected by the water temperature. The incorporation of long-chain fatty acids and the activity of elongases and desaturases are stimulated in the process of

TABLE 3
Optimal Dietary Protein and Lipid for Rapid Growth of
Atlantic Salmon

Fish size	Protein		Lipid	
	g/100 g	g/MJ GE	g/100 g	g/MJ GE
Fry (0.13–2.0 g)[25]	50	21	26	11
Fingerlings (2–50 g)[26]	44	21	—	—
	46[27]		16	
Growers (0.5–10 kg)[28,29]	35–44	17–21	25–27	12

membrane viscosity adaptation to low environmental temperatures.[34] This may also indicate that the water temperature affects the fatty acid requirement. In vitro measurements of desaturase activities indicate that salmon are able to make C20 essential fatty acids from 18:3n-3 and 18:2n-6. The ability to subsequently produce 22:3 and 22:6 fatty acids from the C20 n-3 fatty acids, however, seems limited.[35]

A definite upper limit for fat in practical diets for Atlantic salmon has not been defined. Investigations of the effects of dietary fat content on the growth of Atlantic salmon in fresh and saltwater (Table 3) have shown, though, that fry and growing/finishing fish have the highest growth when fed dry diets containing 25 to 27% fat.[25,28,29] Our unpublished results indicate that this corresponds to an optimal dietary fat content of 11 to 12 g/MJ GE.

CARBOHYDRATES

No published information is available concerning maximum tolerable levels or physiological responses to various types of dietary carbohydrate by Atlantic salmon.

VITAMINS AND MINERALS

The known requirements of Atlantic salmon for micronutrients are compiled in Table 4. As can be seen from this list, only seven have been quantified. This means that diet formulations for Atlantic salmon have to be based largely on extrapolated information from other species.

A list of reported micronutrient deficiency signs in Atlantic salmon is given in Table 5. It is important when formulating diets to be aware not only of possible dietary deficiencies, but also the possibility of toxicity situations. Grisdale-Helland et al.[39] reported a small increase in mortality among juvenile Atlantic salmon that received supplemental vitamin A. The levels of vitamin A used in that experiment were within the normal range that can be expected in fish meal made from blue whiting.[54]

PRACTICAL DIET FORMULATION

The nutritive value of salmon diets is affected by the feedstuff composition and quality. Main factors affecting the quality of the fish meal include fish species, freshness of the raw materials, and the drying conditions.[18,55] In addition, fat source,[31] the source and treatment of the carbohydrate,[19] and the type and content of vegetable feedstuffs containing fiber and antinutrients[56-60] clearly affect the nutritive value of the diet. Because of this multitude of factors influencing the diet, digestibility varies widely.

Because of the difficulty in obtaining suitable quantities of feces from small fish, the digestibility of various dietary components has not been studied and reported in detail for Atlantic salmon weighing less than 1 kg. In studies with Atlantic salmon over 1 kg, the digestibility of crude protein in fish meal has been found to be in the range of 80 to 90%, whereas approximately 91 to 96% of the fat in fish oil was digested.[18,19] The digestibility of

TABLE 4
Summary of Micronutrient Requirements
Established for Atlantic Salmon[a]

Micronutrient	Requirement	Ref.
Vitamin E	35 mg/kg diet	36
Pyridoxine	5 mg/kg diet[b]	37
Vitamin C	50 mg/kg diet[b]	38
Vitamin K	R[c]	39
Riboflavin	R	40
Pantothenic acid	R	40
Phosphorus	1.3% of diet[d]	41
Manganese	20 mg/kg diet	42
Copper	6 mg/kg diet[d]	43
Iron	73 mg/kg diet[d,e]	43
Selenium	R	44–46
Iodine and fluorine	R	47
Potassium	NR	48
Keto-carotenoids	R	49,50

[a] R = required; NR = not required under the experimental conditions used.

[b] No indication was given about the level of the component in the basal diet.

[c] Salmon grew better when vitamin K_1 was supplemented in comparison with menadione sodium bisulfite supplementation.

[d] The level of the component in the basal diet is added to the optimum supplemental level.

[e] Value presented based on tissue accumulation data; value would be 43 mg/kg diet if based on response of hemoglobin and other blood parameters.

starch from extruded wheat has been reported to be 57%, whereas that of gelatinized corn starch was 64%.[14] In general, the digestibility of carbohydrate and other nutrients was negatively affected by high dietary carbohydrate levels.[61]

STARTER DIETS

Farmed Atlantic salmon in the freshwater stage are usually fed dry diets based on special qualities of fish meal, oil, and gelatinized grain products, in addition to vitamins and minerals. Semimoist diets may also be used.[62] The formulation of a typical starter diet is presented in Table 6.

Starter diets have traditionally been crumbled, resulting in a relatively uneven particle size and varying ratios of pelleted and unpelleted material in the final product. Recently, agglomerated diets have been introduced. In this process, powder is spun in the presence of water, which results in granules being formed.[63] The main difference with these diets is that they are uniform in size and can be produced without carbohydrate. For larger fingerlings and smolts, steam-pelleted or extruded diets are used. Recently, betaine has also been added to some starter diets because of its influence on the osmoregulatory capacity of Atlantic salmon parr.[64]

GROWER DIETS

In the sea period, Atlantic salmon are usually fed either moist or dry diets. Moist diets contain binder meal, fish oil, and moist feed ingredients. The binder meal normally contains fish meal, gelatinized grain products, and micronutrients, and may contain some protein from vegetable sources and active binders, such as alginate, carboxymethyl cellulose, and guar

TABLE 5
Major Micronutrient Deficiency Signs in Atlantic Salmon

Micronutrient	Deficiency signs
Pyridoxine[37,51]	Reduced alanine amino transferase activity, degenerative changes in liver, kidney, and gill lamella; nervous disorders; anorexia; poor growth and disintegration of erythrocytes
Vitamin C[38,46]	Increased mortality, slow growth, lethargy, scoliosis, lordosis, broken backs, and anemia
Vitamin E[36,45,52,53]	Reduced growth, increased mortality, skeletal muscle degeneration, myocardial degradation, anemia, impaired serum complement function, elevated plasma proteins, depressed carcass protein and ash content, and increased carcass fat and water content
Iron[43]	Reduced hematocrit; red blood cell count; hemoglobin and serum; liver, spleen, and kidney iron
Copper[43]	Reduced serum copper and liver cytochrome C oxidase activity
Manganese[42]	Reduced hematocrit and hemoglobin and reduced vertebrae Mn
Selenium[45,46]	Lethargy, anorexia, increased mortality, reduced muscle tone and glutathione peroxidase activity
Selenium and vitamin E[45]	Muscular dystrophy

gum. In addition to ensuring a proper feed consistency, the binder meal supplies a major part of the nutrients to the diet. The moist feed ingredients may be various frozen or ensiled products.[32,65-68] Silage made with propionic acid should be avoided in diets for Atlantic salmon, since this compound has negative effects on growth.[68] One advantage of moist diets is that they are readily accepted by the fish. The high acceptability of the moist diets cannot be explained by the high water content,[69] but may be related to their freshness.[70] These feeds are made from local feed resources, such as byproducts of fish processing, which may otherwise be wasted and a source of pollution.

The most common types of feed fed to growing/finishing Atlantic salmon are dry diets processed by steam pelleting or extrusion. Extruded diets can take up more fat than steam pelleted diets and sink more slowly in the water. Some producers base the protein supply on fish meal only, while others use up to 20% vegetable protein feedstuffs, such as soybean meal. Up to now, fish oil has been the only type of oil used in the production of diets for growing and finishing Atlantic salmon in Norway.

SPECIAL DIETS

PIGMENTER DIETS

Atlantic salmon must be fed dietary keto-carotenoids (astaxanthin or canthaxanthin) in order to develop the characteristic pink flesh color. Astaxanthin, the carotenoid found in flesh of wild salmon,[71] is more efficiently utilized[72] and gives a more reddish color to the flesh[73] than canthaxanthin (which imparts a more yellowish color). It is common practice to use 40 to 50 mg carotenoid per kilogram of dry feed for at least 1 year prior to slaughter to obtain satisfactory pigmentation.

TABLE 6
Example of a Steam-Pelleted, Crumbled Starter Diet Formulation for Atlantic Salmon[17]

Ingredient	g/kg diet
Fish meal (Norse LT 94)	650
Blood meal	20
Skimmed milk powder	10
Whey powder	10
Grass meal	10
Seaweed meal	5
Soy lecithin	10
Extruded wheat	135
Fish oil (NorSalmOil)	110
Ground limestone	5
Salt (NaCl)	5
Vitamin and mineral premix[a]	20

Proximate composition (air dry)	
Dry matter	895
Crude protein	476
Crude fat	168
Ash	113
Crude fiber	17
N-free extract[b]	121

[a] Amounts per kg of premix: Vitamin A, 250,000 IU; Vitamin D_3, 50,000 IU; menadione, 1.0 g; α-tocopherol, 60 g; thiamine, 1.0 g; riboflavin, 2.5 g; pyridoxine, 1.5 g; Ca-pantothenate, 4.0 g; niacin, 1.5 g; folic acid, 0.5 g; vitamin B_{12}, 2.0 mg; biotin, 20.0 mg; choline chloride, 60 g; inositol, 10 g; para-aminobenzoic acid, 25 mg; ascorbic acid, 20 g; Fe, 4.0 g ($FeSO_4$); Mn, 4.0 g (MnO); Zn 5.0 g (ZnO); Cu, 0.8 g (CuO), I, 0.15 g [$Ca(IO_2)_3$]; Se, 10 mg (Na_2SeO_4).

[b] Calculated by difference.

BROODSTOCK DIETS

The nutrient requirements of Atlantic salmon broodstock have not been studied in detail. Commercial broodstock diets are normally formulated similar to the grower diets but contain extra vitamins (especially vitamins C and E), and may contain higher levels of carotenoids.

SMOLTIFICATION DIETS

Results have shown that dietary salt[4] and betaine[64] aid in the maintenance of ionic and osmotic balance of Atlantic salmon in connection with smoltification. Special smoltification diets with up to 6% salt (NaCl), in addition to that found in the fish meal, and/or betaine-containing feedstuffs are sold.

MEDICATED DIETS

Upon receipt of a prescription from a veterinarian, feed containing antibiotics can be purchased in Norway for therapeutic reasons. Medicated feed cannot be obtained for use in the prevention of diseases or for possible growth-stimulating effects on Atlantic salmon. Special care is taken in the production of these feeds to prevent binding of the medication to other feed components and to make them palatable to sick fish.[74] A short-term experiment has shown that the presence of 1% oxytetracycline in a moist diet fed to healthy rainbow trout resulted in over a 60% reduction in feed intake.[75]

REFERENCES

1. **USDC/NMFS,** World salmon aquaculture, IFR 90/30, 1990.
2. **Hoar, W. S.,** The physiology of smolting salmonids, in *Fish Physiology*, Vol. 11B, Hoar, W. S. and Randall, D. J., Eds., Academic Press, London, 1988, 275.
3. **McCormick, S. D. and Saunders, R. L.,** Preparatory physiological adaptations for marine life of salmonids: osmoregulation, growth and metabolism, *Am. Fish. Soc. Symp.*, 1, 211, 1987.
4. **Basulto, S.,** Induced saltwater tolerance in connection with inorganic salts in the feeding of Atlantic salmon (*Salmo salar* L.), *Aquaculture*, 8, 45, 1976.
5. **Higgins, P. J.,** Metabolic differences between Atlantic salmon (*Salmo salar*) parr and smolts, *Aquaculture*, 45, 33, 1985.
6. **Aksnes, A., Gjerde, B., and Roald, S. O.,** Biological, chemical and organoleptic changes during maturation of farmed Atlantic salmon, *Salmo salar*, *Aquaculture*, 53, 7, 1986.
7. **Gjedrem, T.,** Genetic variation in quantitative traits and selective breeding in fish and shellfish, *Aquaculture*, 33, 51, 1983.
8. **Fevolden, S. E., Refstie, T., and Røed, K. H.,** Breeding for high and low cortisol stress response in Atlantic salmon (*Salmo salar*) and rainbow trout (*Oncorhynchus mykiss*), *Aquaculture*, submitted.
9. **Saunders, R. L., Henderson, E. B., and Harmon, P. R.,** Effects of photoperiod on juvenile growth and smolting of Atlantic salmon and subsequent survival and growth in sea cages, *Aquaculture*, 45, 55, 1985.
10. **Saunders, R. L. and Henderson, E. B.,** Effects of constant day length on sexual maturation and growth of Atlantic salmon (*Salmo salar*) parr, *Can. J. Fish. Aquat. Sci.*, 45, 60, 1988.
11. **Björnsson, B. T., Thorarensen, H., Hirano, T., Ogasawara, T., and Kristinsson, J. B.,** Photoperiod and temperature affect plasma growth hormone levels, growth, condition factor and hypoosmoregulatory ability of juvenile Atlantic salmon (*Salmo salar*) during parr-smolt transformation, *Aquaculture*, 82, 77, 1989.
12. **Saunders, R. L., Specker, J. L., and Komourdjian, M. P.,** Effects of photoperiod on growth and smolting in juvenile Atlantic salmon (*Salmo salar*), *Aquaculture*, 82, 103, 1989.
13. **Stefansson, S. O., Naevdal, G., and Hansen, T.,** The influence of three unchanging photoperiods on growth and parr-smolt transformation in Atlantic salmon *Salmo salar*, *J. Fish. Biol.*, 35, 237, 1989.
14. **Rumsey, G. L. and Ketola, H. G.,** Amino acid supplementation of casein in diets of Atlantic salmon (*Salmo salar*) fry and of soybean meal for rainbow trout (*Salmo gairdneri*) fingerlings, *J. Fish. Res. Board Can.*, 32, 422, 1975.
15. **Farmer, G. J., Ashfield, D., and Goff, T. R.,** A Feeding Guide for Juvenile Atlantic Salmon, Can. MS Rep. Fish. Aquat. Sci., No. 1718, 1983.
16. **Austreng, E., Storebakken, T., and Åsgård, T.,** Growth rate estimates for cultured Atlantic salmon and rainbow trout, *Aquaculture*, 60, 157, 1987.
17. **Storebakken, T. and Austreng, E.,** Ration levels for salmonids. I. Growth, survival, body composition, and feed conversion in Atlantic salmon fry and fingerlings, *Aquaculture*, 60, 189, 1987.
18. **Andorsdóttir, G.,** Protein Quality, Methionine Supplementation and Fat Levels in Starter Diets for Salmon Fry, Cand. Scient. thesis, University of Oslo, Oslo, 1986.
19. **Rosenlund, G.,** Digestibility Studies with Atlantic Salmon, Tech. Rep., The Foundation for Scientific and Industrial Research at the Norwegian Institute of Technology, No. STF21 A88042, 1988.
20. **Wilson, R. P. and Cowey, C. B.,** Amino acid composition of whole body tissue of rainbow trout and Atlantic salmon, *Aquaculture*, 48, 373, 1985.
21. **Ketola, H. G.,** Amino acid nutrition of fishes: requirements and supplementation of diets, *Comp. Biochem. Physiol.*, 73B, 17, 1982.
22. **Cowey, C. B., Daisley, K. W., and Parry, G.,** Study of amino acids, free or as components of protein, and of some B vitamins in the tissues of the Atlantic salmon, *Salmo salar*, during spawning migration, *Comp. Biochem. Physiol.*, 7, 29, 1962.
23. **Erenst, V.,** Nutrient Contents in Eggs of Atlantic Salmon, *Salmo salar*, Relation with Early Survival and Changes During Development, Int. Rep., Dept. Fish Culture and Inland Fisheries, Agricultural University of Wageningen, 1985.
24. **National Research Council,** *Nutrient Requirements of Coldwater Fishes*, National Academy Press, Washington, D.C., 1981.
25. **Andorsdóttir, G. and Austreng, E.,** Fat level in starter diet for Atlantic salmon, presented at Eur. Soc. Comp. Physiol. Biochem., 7th Conf., Barcelona, Aug. 26–28, 1985.
26. **Austreng, E.,** Fat and protein in diets for salmonid fishes. IV. Protein content in dry diets for salmon parr, *Meld. Norg. LandbrHøgsk.*, 56(19), 1, 1977.
27. **Austreng, E.,** Fat and protein in diets for salmonid fishes. I. Fat content in dry diets for salmon parr, (*Salmo salar*, L.), *Meld. Norg. LandbrHøgsk.*, 55(5), 1, 1976.
28. **Austreng, E. and Storebakken, T.,** Practical formulation of salmonid diets with emphasis on fat and protein, Acts of the Norwegian-French Workshop on Aquaculture, Brest, Dec. 5–8, 1984.

29. **Storebakken, T.,** unpublished data, 1981.
30. **Austreng, E., Grisdale-Helland, B., Helland, S. J., and Storebakken, T.,** Farmed Atlantic salmon and rainbow trout, *Livest. Prod. Sci.,* 19, 369, 1988.
31. **Hardy, R. W., Scott, T. M., and Harell, L. W.,** Replacement of herring oil with menhaden oil, soybean oil, or tallow in the diets of Atlantic salmon raised in marine net-pens, *Aquaculture,* 65, 267, 1987.
32. **Thomassen, M. S. and Røsjø, C.,** Different fats in feed for salmon: influence on sensory parameters, growth rate and fatty acids in muscle and heart, *Aquaculture,* 79, 129, 1989.
33. **Lie, Ø., Waagbø, R., and Sandnes, K.,** Growth and chemical composition of adult Atlantic salmon (*Salmo salar*) fed dry and silage-based diets, *Aquaculture,* 69, 343, 1988.
34. **Bell, M. V., Henderson, R. J., and Sargent, J. R.,** The role of polyunsaturated fatty acids in fish, *Comp. Biochem. Physiol.,* 83B, 711, 1986.
35. **Tocher, D. R. and Dick, J. R.,** Polyunsaturated fatty acid metabolism in cultured fish cells: incorporation and metabolism of (n-3) and (n-6) series acids by Atlantic salmon (*Salmo salar*) cells, *Fish Physiol. Biochem.,* 8, 311, 1990.
36. **Lall, S. P., Olivier, G., Hines, J. A., and Ferguson, H. W.,** The role of vitamin E in nutrition and immune response of Atlantic salmon (*Salmo salar*), *Bull. Aqua. Assoc. Can.,* 88-2, 76, 1988.
37. **Lall, S. P. and Weerakoon, D. E. M.,** Vitamin B_6 requirement of Atlantic salmon (*Salmo salar*), *FASEB J.,* 4 (Abstr.), 3749, 1990.
38. **Lall, S. P., Oliver, G., Weerakoon, D. E. M., and Hines, J. A.,** The effect of vitamin C deficiency and excess on immune response in Atlantic salmon, in *The Current Status of Fish Nutrition in Aquaculture,* Takeda, M. and Watanabe, T., Eds., Tokyo University of Fisheries, Tokyo, 1990, 427.
39. **Grisdale-Helland, B., Helland, S. J., and Åsgård, T.,** Problems associated with the present use of menadione sodium bisulfite and vitamin A in diets for Atlantic salmon, *Aquaculture,* 92, 351, 1991.
40. **Phillips, A. M., Jr., Podoliak, H. A., Dumas, R. F., and Thoesen, R. W.,** Vitamin requirement of Atlantic salmon, *Fish. Res. Bull., N.Y.,* 22, 79, 1958.
41. **Ketola, H. G.,** Requirement of Atlantic salmon for dietary phosphorus, *Trans. Am. Fish. Soc.,* 104, 548, 1975.
42. **Lall, S. P. and Hines, J. A.,** Manganese bioavailability and requirements of Atlantic salmon (*Salmo salar*) and brook trout (*Salvelinus fontinalis*), XIII Int. Cong. Nutr., Sat. Symp., Brighton, Aug. 17–18, 1985 (Abstr.).
43. **Lall, S. P. and Hines, J. A.,** Iron and copper requirements of Atlantic salmon (*Salmo salar*), Int. Symp. Feeding and Nutrition in Fish, Bergen, Aug. 23–27, 1987 (Abstr.).
44. **Bell, J. G., Cowey, C. B., Adron, J. W., and Pirie, B. J. S.,** Some effects of selenium deficiency on enzyme activities and indices of tissue peroxidation in Atlantic salmon parr (*Salmo salar*), *Aquaculture,* 65, 43, 1987.
45. **Poston, H. A., Combs, G. F., Jr., and Leibovitz, L.,** Vitamin E and selenium interrelationships in the diet of Atlantic salmon (*Salmo salar*): gross, histological and biochemical deficiency signs, *J. Nutr.,* 106, 892, 1976.
46. **Poston, H. A. and Combs, G. F., Jr.,** Interrelationships between requirements for dietary selenium, vitamin E, and L- ascorbic acid by Atlantic salmon (*Salmo salar*) fed a semipurified diet, *Fish Health News,* 8 (4), 6, 1979.
47. **Lall, S. P., Paterson, W. D., Hines, J. A., and Adams, N. J.,** Control of bacterial kidney disease in Atlantic salmon, *Salmo salar* L., by dietary modification, *J. Fish Dis.,* 8, 113, 1985.
48. **Austic, R. E., Jiao, R. R., Rumsey, G. L., and Ketola, H. G.,** Monovalent minerals in the nutrition of salmonid fishes, Proc. Cornell Nutr. Conf., East Syracuse, NY, Oct. 24–26, 1989, 25.
49. **Weedon, B. C. L.,** Occurrence, in *Carotenoids,* Isler, O., Ed., Birkhauser Verlag, Basel, 1971, 30.
50. **Torrissen, O. J.,** Pigmentation of salmonids — effect of carotenoids in eggs and start-feeding diet on survival and growth rate, *Aquaculture,* 43, 185, 1984.
51. **Herman, R. L.,** Histopatalogy associated with pyridoxine deficiency in Atlantic salmon (*Salmo salar*), *Aquaculture,* 46, 173, 1985.
52. **Lall, S. P., Olivier, G., Hines, J. A., and Ferguson, H. W.,** The role of vitamin E in Atlantic salmon (*Salmo salar*) nutrition and immune response, Int. Symp. Feeding and Nutrition in Fish, Bergen, Aug. 23–27, 1987 (Abstr.).
53. **Hardie, L. J., Fletcher, T. C., and Secombes, C. J.,** The effect of vitamin E on the immune response of the Atlantic salmon (*Salmo salar* L.), *Aquaculture,* 87, 1, 1990.
54. **Asbjørnsen, B.,** Vitamin A- og E-innhold i fiskemel fra forskjellige råstoffslag, Meldinger fra Sildolje- og Sildemelindustriens Forskningsinstitutt, 2-88, 17, 1988.
55. **Åsgård, T.,** Nutritional value of animal protein sources for salmonids, in *Proc. Aquaculture International Congress & Exposition,* Vancouver, BC, 1988, 411.
56. **Arnesen, P., Brattås, L.-E., Olli, J., and Krogdahl, Å.,** Soybean carbohydrates appear to restrict the utilization of nutrients by Atlantic salmon (*Salmo salar* L.), in *The Current Status of Fish Nutrition in Aquaculture,* Takeda, M. and Watanabe, T., Eds., Tokyo University of Fisheries, Tokyo, 1990, 273.
57. **Krogdahl, Å.,** Alternative protein sources from plants contain antinutrients affecting digestion in salmonids, in *The Current Status of Fish Nutrition in Aquaculture,* Takeda, M. and Watanabe, T., Eds., Tokyo University of Fisheries, Tokyo, 1990, 253.

58. **Krogdahl, Å.,** Dietary fibre affects intestinal conditions of salmonids, Proc. NJF-seminar, Plant Carbohydrates and Associated Components, Herning, Denmark, June 18–20, 1990.

59. **Hendrix, H. G. C. J. M., van den Ingh, T. S. G. A. M., Krogdahl, Å., Olli, J. J., and Koninkx, J. F. J. G.,** Binding of soybean agglutinin to small intestinal brush border membranes and brush border membrane enzyme activities in Atlantic salmon (*Salmo salar*), *Aquaculture*, 91, 163, 1990.

60. **van den Ingh, T. S. G. A. M., Krogdahl, Å., Olli, J. J., Hendrix, H. G. C. J. M., and Koninkx, J. F. J. G.,** Effects of soybean containing diets on the proximal and distal intestine in Atlantic salmon (*Salmo salar*): a morphological study, *Aquaculture*, in press.

61. **Krogdahl, Å.,** personal communication, 1990.

62. **Lemm, C. A.,** Growth and survival of Atlantic salmon fed semimoist or dry starter diets, *Prog. Fish-Cult.*, 45, 72, 1983.

63. **Thorsen, F.,** personal communication, 1990.

64. **Virtanen, E., Junnila, M., and Soivio, A.,** Effects of food containing betaine/amino acid additive on the osmotic adaptation of young Atlantic salmon, *Salmo salar* L., *Aquaculture*, 83, 109, 1989.

65. **Åsgård, T. and Austreng, E.,** Casein silage as feed for salmonids, *Aquaculture*, 48, 233, 1985.

66. **Åsgård, T. and Austreng, E.,** Dogfish offal, ensiled or frozen, as feed for salmonids, *Aquaculture*, 49, 289, 1985.

67. **Åsgård, T. and Austreng, E.,** Blood, ensiled or frozen, as feed for salmonids, *Aquaculture*, 55, 263, 1986.

68. **Austreng, E. and Åsgård, T.,** Fish silage and its use, in *Trend and Problems in Aquaculture Development*, Proc. 2nd Intern. Conf. Aquafarming, Verona, Oct. 12–13, 1984, 218.

69. **Hughes, S. G.,** Effect of dietary moisture level on response to diet by Atlantic salmon, *Prog. Fish-Cult.*, 51, 20, 1989.

70. **Åsgård, T.,** personal communication, 1990.

71. **Khare, A., Moss, G. P., Weedon, B. C. L., and Matthews, A. D.,** Identification of astaxanthin in Scottish salmon, *Comp. Biochem. Physiol.*, 45B, 971, 1973.

72. **Storebakken, T., Foss, P., Schiedt, K., Austreng, E., Liaaen-Jensen, S., and Manz, U.,** Carotenoids in diets for salmonids. IV. Pigmentation of Atlantic salmon with astaxanthin, astaxanthin dipalmitate and canthaxanthin, *Aquaculture*, 65, 279, 1987.

73. **Skrede, G. and Storebakken, T.,** Characteristics of color in raw, baked and smoked wild and pen-reared Atlantic salmon, *J. Food Sci.*, 51, 804, 1986.

74. **Hole, R.,** personal communication, 1990.

75. **Hustvedt, S. O., Storebakken, T., and Salte, R.,** Does oral administration of oxolinic acid or oxytetracylin affect feed intake of rainbow trout? *Aquaculture*, 92, 109, 1991.

AYU, *PLECOGLOSSUS ALTIVELIS*

Akio Kanazawa

INTRODUCTION

Ayu are an anadromous fish that spend most of their 1-year life cycle in fresh water. After spending their first few months of life in the sea, 20 to 30 mm fry enter rivers, feeding on benthic organisms such as diatoms and blue-green algae. After they have reached lengths of up to 25 cm, spawning occurs and the newly hatched fry migrate to sea.

Ayu have been cultured for many years in Japan because of their high demand as a food fish. In Japan, the total production has increased from 3200 t in 1970 to 13,000 t in 1988. Since the successful development of artificial spawning and larval rearing in 1932, rotifers, *Daphnia*, and *Artemia* have been used as larval food for mass production of ayu fry. Recently, microparticulate diets have been developed as a substitute for live food.[1,2] Most cultured ayu are fed artificially processed feed from total body lengths of 40 to 50 mm up to marketable size. Annual feed production for ayu in Japan was 24,500 t in 1988.

NUTRIENT REQUIREMENTS

AMINO ACIDS AND PROTEIN

Fish, such as carp, do not efficiently utilize free dietary amino acids. Thus, the incorporation of radioactive acetate into individual amino acids was investigated in ayu in order to determine the qualitative essential amino acids. Twelve days after an injection of 200 μCi [U-^{14}C]acetate in 2.2 g ayu, the muscle and hepatopancreas were removed. After protein separation and hydrolysis, the radioactivity of each individual amino acid was measured using HPLC and a radioanalyzer. Based on the radioactive incorporation data, it was concluded that the essential amino acids for the ayu are as follows: arginine, methionine, valine, lysine, threonine, isoleucine, leucine, histidine, phenylalanine, and tryptophan.[3]

The optimum dietary protein level for ayu has been shown to be 45 to 50%. Generally, proteins having an essential amino acid profile similar to that of the whole body protein of the fish are likely to have higher nutritive value. Therefore, the amino acid composition of whole-body protein of ayu was analyzed (Table 1). Various protein sources to simulate the amino acid profile of the body protein of ayu were chosen for diet formulation (Table 2). Diets containing three protein levels were compared with a live food (control) diet in a feeding trial using 10-day-old ayu fry. The results of the 30-d feeding trial are shown in Figure 1. The optimum protein level in a microparticulate diet was 40% when the amino acid profile of the diet simulated that of the body protein of larval ayu.[4]

FATTY ACIDS

Since most fish are incapable of *de novo* synthesis of 18:2n-6, 18:3n-3, 20:5n-3, and 22:6n-3 fatty acids, dietary sources of these fatty acids are likely to be essential for normal growth and survival. The effect of essential fatty acids on growth of ayu is shown in Figure 2.[5] The efficacy of 18:2n-3 as an essential fatty acid was similar to that of 20:5n-3 in ayu as well as in rainbow trout and chum salmon.[6] This means that 18:3n-3 is converted to 20:5n-3 in the ayu.[7] The requirement of ayu for 18:3n-3 or 20:5n-3 has been estimated to be about 1.0% of the diet.[5]

TABLE 1
Relative Ratio of Essential Amino Acids to
Methionine in Ayu and Test Diets in Table 2

Amino acid	Ayu	Diets 1, 2, and 3
Methionine	1.00	1.00
Threonine	1.18	1.15
Valine	1.46	1.56
Isoleucine	1.33	1.34
Leucine	2.39	2.35
Phenylalanine	1.36	1.40
Histidine	0.71	0.74
Lysine	2.70	2.31
Tryptophan	0.89	0.84
Arginine	2.08	2.14

TABLE 2
Composition of Test Diets for Ayu
Expressed as g/100 g Dry Diet

Ingredient	Diet 1	Diet 2	Diet 3
Bonito egg	15.62	17.85	20.08
Squid meal	5.92	6.77	7.62
Krill meal	4.93	5.63	6.33
White fish meal	19.13	21.86	24.59
Soybean protein	8.46	9.67	10.88
Gluten meal	3.60	4.11	4.62
Beer yeast	4.11	4.70	5.29
Soybean lecithin	3.00	3.00	3.00
Pollack liver oil	6.00	6.00	6.00
Vitamin mixture	6.00	6.00	6.00
Mineral mixture	5.00	5.00	5.00
Cellulose	18.23	9.41	0.59
Protein content (%)	35.00	40.00	45.00

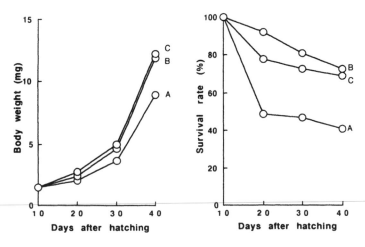

FIGURE 1. Effect of protein level on growth and survival of ayu larvae. (A) Diet 1, 35%; (B) diet 2, 40%; (C) diet 3, 45% protein.

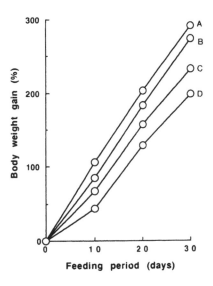

FIGURE 2. Effect of essential fatty acids on growth of ayu. (A) 8% 18:1n-9 + 1% 20:5n-3; (B) 8% 18:1n-9 + 1% 18:3n-3; (C) 8% 18:1n-9 + 1% 18:2n-6; (D) 9% 18:1n-9. (From Kanazawa, A. et al., *Nippon Suisan Gakkaishi*, 48, 588, 1982. With permission.)

TABLE 3
Growth and Survival of Ayu Larvae Fed Test Diets
Containing Various Phospholipids[10]

Dietary lipid	Body weight (mg)	Total length (mm)	Weight/ length ratio	Weight gain (%)	Survival (%)
9.0% Pollack liver oil (PLO)	542	48.3	11.2	442	67
7.5% PLO + 1.5% dipalmitoyl PC	578	47.4	12.2	478	7
7.5% PLO + 1.5% soybean PI	849	50.0	16.8	749	70
7.5% PLO + 1.5% soybean PC	702	51.4	13.6	602	87
7.5% PLO + 1.5% chicken-egg PC	605	53.0	11.4	505	77

From Kanazawa, A. et al., *Z. Angew. Ichthyol.*, 4, 168, 1985. With permission.

PHOSPHOLIPIDS

During the course of investigating the nutritional requirements of larval fish, we observed that dietary sources of phospholipids were essential for normal growth and survival of fish larvae such as the ayu. The addition of some phospholipids to microparticulate diets markedly improved both growth and survival of the larval fish.

Kanazawa et al.[8,9] have investigated the phospholipid requirements of 10-, 30-, and 100-day-old larval ayu. Furthermore, Kanazawa et al.[10] have examined the supplemental effects of highly purified commercial phospholipids on larval ayu (Table 3). Soybean phosphatidyl-

TABLE 4
Vitamin Deficiencies Signs Observed in Ayu[11]

Vitamin	Deficiency signs
Thiamine	Loss of appetite; poor growth; ataxia; high mortality; loss of sense of distance; dullness; cataracts; hemorrhages of the fin bases; congested head; convulsions before death
Riboflavin	Loss of appetite; poor growth; high mortality; hemorrhagic eyes; erosion, hemorrhage and ulceration of the skin on the fin behind and upper part of lateral side
Pyridoxine	Loss of appetite; poor growth; high mortality; hemorrhage; dark color; mild exophthalmia; abdominal edema; erratic swimming
Niacin	Loss of appetite; poor growth; erratic swimming; epithelial and fin base hemorrhages; deformed gill operculum
Pantothenic acid	Loss of appetite; poor growth; ataxia; high mortality; exophthalmia; hemorrhagic eyes; fin and epithelial hemorrhages; congested head
Biotin	None detected
Inositol	None detected
Choline	None detected
Folic acid	Epithelial and fin hemorrhages; erosion of mucous tissues; skin lesions
Ascorbic acid	Loss of appetite; mild exophthalmia; eyes and fin base hemorrhages; congestion of the back of the head; gill operculum and lower jaw erosions

From Aoe, H., in *Nutrition and Diets in Fish,* Ogino, C., Ed., Koseisha-Koseikaku, Tokyo, 1980, 186. With permission.

choline (purity 98%), chicken-egg phosphatidylcholine (purity 98%), and soybean phosphatidylinositol (purity 70%) were found to greatly improve the survival rate, whereas synthetic phosphatidylcholine dipalmitoyl was ineffective. Regarding weight gain, soybean phosphatidylinositol showed the greatest activity, followed by soybean phosphatidylcholine and chicken-egg phosphatidylcholine, respectively. These compounds, in addition to n-3 highly unsaturated fatty acids, are believed to be indispensable for normal growth and survival of larval ayu.

The incidence of malformations, such as protrusion of the thorax, scoliosis, and twisting of the caudal peduncle, is a serious problem in fry production of ayu in Japan. Scoliosis reduces the market value of ayu remarkably. Kanazawa et al.[8] have found that the incidence of malformations, particularly scoliosis, can be reduced by the addition of phospholipids in the diets.

VITAMINS

Water-soluble vitamins that are essential in diets for ayu and the gross deficiency signs of each are summarized in Table 4. Ascorbic acid is quite liable in diets due to oxidation during feed processing and subsequent storage, and may be leached into the water before being eaten by the fish. Recently, the utilization of stable vitamin C compounds such as L-ascorbyl-2-phosphate·Mg and L-ascorbyl-6-stearate were compared in feeding trials using ayu, 60 d of age and 30 mm in total length. Ayu showed better growth when fed diets containing L-ascorbyl-2-phosphate·Mg than with L-ascorbyl-6-stearate or L-ascorbate·Na.[12]

TABLE 5
Relative Composition of an Expanded Pellet Diet for Ayu[14]

Classification of material	Percentage composition	Materials
Animal feed meals	53–57	Fish meal, krill meal, meat meal
Cereals	17–30	Wheat flour, rice powder, bread powder, dextrin
Plant oil cakes	7–11	Soybean meal (defatted), corn gluten meal
Brans	6–16	Rice bran, defatted rice bran
Others	4–10	Wheat germ, dried yeast, plant oils, $CaHPO_4$, $CaCO_3$, NaCl

From *Fish Culture*, 23(12), 111, 1986. With permission.

The requirement of ayu for fat-soluble vitamins has also been demonstrated. Growth and survival were extremely low when ayu were fed diets deficient in β-carotene, calciferol, α-tocopherol, or menadione when compared to fish fed diets with a complete fat-soluble vitamin mixture. Deficiency signs, such as poor appetite, loss of vitality, white color, exophthalmus, spinal curvature, and convulsions before death were observed in ayu fed fat-soluble-vitamin deficient diets.

PIGMENTS

It is necessary to restore the skin color, since cultured ayu exhibit a darker coloration compared with that of wild ayu. The color of ayu originates from zeaxanthin that accumulates in the skin. Wild ayu obtain zeaxanthin from their natural food, i.e., diatoms and blue-green algae. The color of cultured ayu is improved by dietary supplementation of 3 g of *Spirulina* (containing a large amount of zeaxanthin) and 4 mg of lutein per 100 g of diet.[13]

PRACTICAL DIET FORMULATION

Pellet crumbles had previously been used as the major practical diet form for ayu. However, crumbles produced from expanded pellets were developed about 6 years ago, and the use of the expanded pellet produced by extruders has become widespread as the major practical diet form for ayu. The current high-density culture of ayu was made possible by the development of the expanded pellet. These diets provide high growth rates, feed efficiency, and high nutrient digestibility. The composition of an expanded pellet diet for ayu is shown in Table 5. The recommended levels of vitamins to be used in practical ayu feeds are presented in Table 6.

MICROPARTICULATE DIET

In the seed production of aquatic animals for aquaculture, live food, such as rotifers and *Artemia*, have been widely used throughout the world. The seed production of ayu was first achieved by partial, or even in some cases, complete substitution of a microparticulate diet for the live food.[16]

Ayu larvae, numbering 100,000 and 400,000, were placed in 50- and 100-ton tanks, respectively. Feeding trials were carried out for 83 d at 11.1 to 27.0°C. Larvae received the microparticulate diet 10 times daily. The microparticulate diet was formulated as shown in Table 7. One group was fed mostly live food, such as rotifers and *Artemia*, and another group was fed mostly the microparticulate diet in combination with 50% rotifers and 20% *Artemia*. The results of growth and survival of larval ayu reared for 83 d are shown in Figure 3. The

TABLE 6
Recommended Vitamin Supplementation
Levels for Practical Ayu Diets[15]

Vitamin	mg/100 g diet
Thiamine	1.2
Riboflavin	4.0
Pyridoxine	1.2
Niacin	10.0
Pantothenate Ca	5.0
Biotin	0.03
Inositol	40.0
Choline chloride	35.0
Folic acid	0.3
Cyanocobalamine	0.002
Ascorbate Ca	30.0
Vitamin A	1000.0 IU
Vitamin D	200.0 IU
Vitamin E	10.0
Vitamin K	1.0

TABLE 7
Composition of Microparticulate Diet
for Ayu Larvae

Ingredient	g/100 g dry diet
Egg yolk	24
Tapes extract	8
Yeast powder	10
Milk casein	10
Egg albumin	15
Bonito extract	5
Amino acid mixture	5
Mineral mixture	6
Vitamin mixture	8
Squid liver oil	4
Soybean lecithin	3
Soybean oil	2

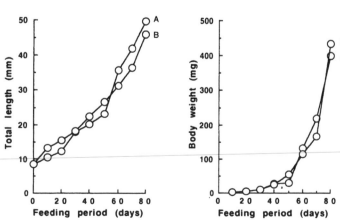

FIGURE 3. Growth of larval ayu fed on microparticulate and live food diets.
(A) Microparticulate diet; (B) live food diet.

survival of larval ayu fed the microparticulate diet and those fed the live food at the end of the feeding trial was 44 and 54%, respectively. Total lengths of the fish fed the microparticulate diet and live food were 50 and 47 mm, respectively. These results indicate that the microparticulate diet has a nutritive value similar to that of the live food in terms of supporting the growth and survival rate.[17]

REFERENCES

1. **Kanazawa, A.,** Microparticulate diets, in *Fish Nutrition and Diets, Suisangaku Ser. 54*, Yone, Y., Ed., Koseisha-Koseikaku, Tokyo, 1985, 99.
2. **Oka, A., Sato, T., and Sarado, F.,** Rearing of larval ayu with artificial microdiets, *The Aquaculture*, 34, 15, 1986.
3. **Kanazawa, A.,** unpublished data, 1987.
4. **Kanazawa, A.,** unpublished data, 1990.
5. **Kanazawa, A., Teshima, S., and Sakamoto, M.,** Requirements of essential fatty acids for the larval ayu, *Nippon Suisan Gakkaishi*, 48, 587, 1982.
6. **Kanazawa, A.,** Essential fatty acid and lipid requirement of fish, in *Nutrition and Feeding in Fish*, Cowey, C. B., Mackie, A. M., and Bell, J. G., Eds., Academic Press, London, 1985, 281.
7. **Kanazawa, A., Teshima, S., and Ono, K.,** Relationship between essential fatty acid requirements of aquatic animals and the capacity for bioconversion of linolenic acid to highly unsaturated fatty acid, *Comp. Biochem. Physiol.*, 63B, 295, 1979.
8. **Kanazawa, A., Teshima, S., Inamori, S., Iwashita, T., and Nagao, A.,** Effects of phospholipids on growth, survival rate, and incidence of malformation in the larval ayu, *Mem. Fac. Fish., Kagoshima Univ.*, 30, 301, 1981.
9. **Kanazawa, A., Teshima, S. Kobayashi, T., Takae, M., Iwashita, T., and Uehara, R.,** Necessity of dietary phospholipids for growth of the larval ayu, *Mem. Fac. Fish., Kagoshima Univ.*, 32, 115, 1983.
10. **Kanazawa, A., Teshima, S., and Sakamoto, M.,** Effects of dietary bonito-egg phospholipids and some phospholipids on growth and survival of the larval ayu, *Plecoglossus altivelis*, *Z. Angew. Ichthyol.*, 4, 165, 1985.
11. **Aoc, H.,** Vitamins, in *Nutrition and Diets in Fish*, Ogino, C., Ed., Koseisha-Koseikaku, Tokyo, 1980, 186.
12. **Kanazawa, A.,** unpublished data, 1990.
13. **Mori, T., Muranaka, T., Miki, W., Yamaguchi, K., Konosu, S., and Watanabe, T.,** Pigmentation of cultured sweet smelt fed diets supplemented with a blue-green alga *Spirulina maxima*, *Nippon Suisan Gakkaishi*, 53, 433, 1987.
14. **Miyazono, I.,** Characteristic and feeding method of expanded pellet feed for ayu, *Fish Culture*, 23, 110, 1986.
15. **Takamatsu, C.,** Fish feeds, in *Lecture on Formulated Feed*, Editorial Committee, Ed., Chikusan Shuppansha, Tokyo, 1980, 579.
16. **Kanazawa, A., Teshima, S., Inamori, S., Sumida, S., and Iwashita, T.,** Rearing of larval red sea bream and ayu with artificial diets, *Mem. Fac. Fish., Kagoshima Univ.*, 31, 185, 1982.
17. **Kanazawa, A. and Teshima, S.,** Microparticulate diets for fish larvae, in *New and Innovative Advances in Biology/Engineering with Potential for Use in Aquaculture*, Sparks, A. K., Ed., NOAA Tech. Rep. NMFS 70, Natl. Mar. Fish. Serv., 1988, 57.

BRAZILIAN FINFISH, TAMBAQUI, PACU, AND MATRINXÃ

Newton Castagnolli

INTRODUCTION

Actual production of Brazil's native species far exceeds 40 million fingerlings due to increased production by private hatcheries.

There is no doubt that Brazilian fish culture is just starting to develop with the growing production of native precocious herbivorous-omnivorous fishes such as tambaqui, *Colossoma macropomum*, pacu, *Piaractus mesopotamicus* (formerly *Colossoma mitrei*), and matrinxã, *Brycon melanopteron*.

Although fish farmers have already started to rear pacu and tambaqui on a commercial basis, most of the nutritional research related to fry, fingerling, and juvenile up to marketable size fish still remains to be done. It may be stated that such knowledge will contribute to the future development of the still incipient fish-farming activity to an industrial scale in Brazil.

NUTRIENT REQUIREMENTS

PROTEIN

The first nutritional study concerning a native fish species was conducted at the fish culture section of UNESP, Campus de Jaboticabal.[1] This study involved the estimation of the protein requirement of tambaqui. The results indicated that 22% crude protein was adequate in the early fingerling stage (5 to 30 g) and 18% crude protein was adequate for the juvenile stage in rearing ponds. These data indicated that this species could be cultured under intensive management systems with artificial diets at lower feed costs than other species that have a higher protein requirement. Protein digestibility values for tambaqui fed diets containing 14, 18, 22, and 26% crude protein were found to be 67.7, 85.9, 81.7, and 74.9%, respectively.[2]

The protein requirement for matrinxã has been estimated to be 35.5%.[3] Protein digestibility studies have been conducted with this species fed 35.5% crude protein diets containing either all plant protein sources or up to 80% from animal protein sources. Digestibility coefficients were very high, above 90%, for all treatments, which indicates that this species can readily digest protein from both plant and animal sources.[4] Thus, plant protein sources should be chosen for this species because they are less expensive and more widely available for fish culture within the Amazonian region. The nutrient content of some amazonian byproducts, such as barley bran, distiller's yeast (*Saccharomyces cereviseae*), and several macrophytes, have been analyzed that could possibly be used in the formulation of feeds for fish culture in that region where there is a serious lack of agroindustrial byproducts. A study with matrinxã fed rations formulated with a mixture of these natural and agroindustrial byproducts produced very good growth performance of the fish.[5]

Dietary crude protein levels of 26 to 30% have been shown to provide the best growth performance of pacu fed isocaloric diets containing 3200 kcal ME/kg diet.[6-9] Digestibility coefficients of about 87% were observed for these diets at the 26 to 30% crude protein level. These workers also estimated the optimum dietary energy level to be 3200 kcal ME/kg diet based on growth data obtained by feeding isonitrogenous diets containing 23% crude protein. Isocaloric (2600 kcal/kg) and isonitrogenous (25% crude protein) diets formulated with different plant and animal protein ratios (100, 85, 70, and 55% plant protein) were fed to pacu without resulting in any significant difference in growth performance.[10] These data indicate that pacu, as well as matrinxã, can utilize plant protein as well as animal protein.

Experiments have also been conducted to study the influence of plant and animal protein sources on the gastrointestinal transit time in pacu as influenced by water temperature.[6] Three water temperatures (24, 28, and 32°C) and two dietary energy levels (3600 and 4000 kcal GE/kg diet) were studied. Temperature had a marked influence on fish growth, feed conversion rate, and protein digestibility, with the better performance observed at the higher temperature. The transit time of the food was also highly influenced by the water temperature, i.e., 12 h at 28 and 32°C and 36 h at 24°C. This observation helped provide recommendations on feeding frequency in culture ponds relative to water temperature. In addition, better performance was observed in fish fed diets containing the higher protein and lower energy content at the higher water temperatures.

Brenner[11] has estimated the dietary protein requirement of pacu using semipurified diets. Diets were formulated with the following ingredients: casein, dextrin, soybean oil, carboxymethyl cellulose, calcium carbonate, dicalcium phosphate, cellulose, vitamin mixture, and mineral mixture. Six dietary crude protein levels ranging from 12 to 52% in isocaloric diets containing 3300 kcal/kg diet were tested. The protein requirement for 22.2 to 36.4 g fish was estimated to be 36% based on growth, protein conversion ratio, and protein retention data. The data also showed that the protein retention ratio was higher when fish were fed the lower protein diets and that feed consumption decreased as dietary protein level increased. The protein requirement value estimated in this study with semipurified diets is somewhat higher than that previously reported using practical diets.[6-9] Higher dietary protein levels (40%) have been shown to improve growth performance in pacu during the winter season.[12]

Carneiro[13] has compared the performance of pacu fed the same diet containing 3200 kcal DE/kg and 26% crude protein processed and fed as a powder, pelleted, and extruded feed. Preliminary results indicate a higher growth rate with lower conversion efficiency for the pacu receiving the extruded feed.

Several studies have been conducted with different management systems involving cages and enclosures, polyculture, integrated systems with swine and poultry, and different stocking densities of pacu and tambaqui. Good production, in excess of 13 tons/ha/year, have been obtained with tambaqui in polyculture with carp and/or tilapia in fertilized earthen ponds with relatively low water inflows. Supplemental feeding with low-protein poultry feeds or some local byproducts, such as coconut meal, were used. This relative high productivity was obtained in the northeastern region of Brazil. Water temperatures in this region provide for an all-year growing season, and the natural productivity of the ponds is high due to the relatively high water alkalinity.

LIPIDS

Pezzato[14] has evaluated various levels and sources of dietary fat in practical diets for pacu. Three levels of fat (8, 16, and 24%) as either plant (soybean oil) or animal (swine fat) origin were supplemented to a diet containing corn, fish meal, soybean meal, vitamin, and mineral premixes. The additional fat was added by replacing corn meal. The experiment was conducted in aquaria for 105 d. The results indicated that pacu can utilize high levels of dietary fat from either plant or animal origin. The higher dietary fat intake resulted in higher body fat levels. Whole-body fat content increased 38.1 and 27.6% in fish fed the animal and plant lipid sources, respectively. No adverse pathological effects were observed in the fish fed the high-lipid diets. Analysis of dietary and tissue fatty acids indicated that pacu are able to convert short-chain fatty acids into long-chain fatty acids.

FEEDING AND NUTRITION OF PACU FRY

Now that induced breeding of South American native migratory fish species has become routine in experimental stations and even in private hatcheries in Brazil, survival of larvae to the fry or fingerling stage remains the major obstacle to fish culture development. Although fingerling production of promising indigenous fish species has increased considerably in Brazil in the last 2–3 years, there is still a need for further research aimed at improving the survival rate. The main constraint is the need for adequate feeding in the early life stage soon after the absorption of the yolk sac. In addition, techniques are needed to prevent predation by dragonfly larvae and other insects.

Although the nutrient requirements for the main South American native species and the techniques of mass production of planktonic microorganisms have not yet been established, semiintensive rearing of fry in floating net-nursery ponds during their first 2 weeks of life seems the best way to achieve reasonable survival rates.

The first pacu larvae produced at our facilities were reared in net-nursery pens floating in recently fertilized ponds.[15] Production of these larvae also permitted the study of their growth during the first month of life, as well as analysis of the planktonic feed organisms taken during the period between the fry and fingerling stage. This work showed that pacu fry fed on rotifers from the fourth day of life, and on the 10th day started to feed on young copepods and nuplii of cladocerans. From the 19th day, the main food items found in the stomach contents were microcrustaceans and chironomid larvae.[16] The production of planktonic microorganisms, such as algae, rotifers, and microcrustacea, for larval feeding has recently been reviewed elsewhere.[17]

Stefani and Castagnolli[18] have tested two different organic fertilizers (chicken and swine manure) combined with two different feed ingredients (rice bran and distiller's yeast) in outdoor brick-layered ponds for the production of the cladoceran *Diaphanosoma*. The best production in all treatments was at temperatures between 27 and 30°C, dissolved oxygen levels of 7 to 9 ppm, and at pH levels between 7.5 and 8.0. The relatively low productivity achieved (127 to 144 organisms/l) was attributed to diurnal temperature oscillations.

Yamanaka[19] has reviewed the development and natural feeding habits of pacu fry, soon after eclosion, up to the prejuvenile (55 mm standard length) stage. He also described the biometric index and meristic characteristics of fry submitted to different rearing techniques considering natural vs. artificial food and to different stocking densities and mesh sizes of the floating net nurseries set in outdoor fertilized ponds. This work also describes the evolution of the digestive tract from fry up to the fingerling or prejuvenile stage.

Sipauba Tavares[20] carried out studies to develop and utilize techniques for large-scale production of algae and zooplancton organisms. The effectiveness of these natural food organisms as promoters of growth of fish larvae and fry was also studied. The author cultured three species of Chlorophycean algae: *Ankistrodesmus gracilis*, *Scenedesmus bijugatus*, and *Chlamidomonas* spp. in CHu nutrient medium compared with a less expensive medium prepared with chemical fertilizer. The algae and zooplankton cultured were utilized as food in experiments on feeding, food selection, survival, and growth of fry. It was pointed out that certain species of algae and zooplankton can be successfully grown under laboratory conditions, and this seems to be a way to reduce the high fish mortality that usually occurs in the hatcheries due to a shortage of adequate food during the first days of life.

Dias[21] stocked 100 l cement aquaria with 14 one-week-old larvae at a constant water flow of 2 to 3 l/min. The larvae were fed twice a day (morning and afternoon) with an artificial diet (30% crude protein) or natural feed (wild plankton microorganisms and/or *Artemia salina*), and a third treatment consisting of a mixture of both. The mixed diet produced the best growth performance. No significant difference was observed in mortality after 45 or 90 d. The average survival rate of all treatments was 50%.

REFERENCES

1. **Macedo, E. M., Carneiro, D. J., and Castagnolli, N.,** Necessida des protéicas na nutrição do tambaqui *Colossoma macropomum* Cuvier, 1818 (Pisces, Characidae), in *Proc. Simp. Bras. Aqüicul. 2*, 2 ENAR Sudepe, Brasília, D.F., 1981, 77.

2. **Carneiro, D. J.,** Digestibilidade protéica em dietas isocalóricas para o tambaqui *Colossoma macropomum* Cuvier (Pisces, Characidae), in *Proc. Simp. Bras. Aqüicul. 2*, ENAR Sudepe, Brasília, D.F., 1981, 78.

3. **Werder, U. and Saint Paul, U.,** Feeding trial with herbivorous and omnivorous Amazonian fishes, *Aquaculture*, 15, 175, 1978.

4. **Cyrino, J. E. P., Castagnolli, N, and Pereira Filho, M.,** Digestibilidade da proteína de origem animal e vegetal pelo matrinxã (*Brycon cephalus*), in *Proc. Simp. Bras. Aqüicul. 4*, Cuiabá, M.T., 1988, 49.

5. **Pereira Filho, M.,** Preparo e Utilização de Ingredientes Produzidos em Manaus, no Arraçoamento de Matrinxã, *Brycon* sp., M.S. thesis, INPA, Manaus, 1987.

6. **Carneiro, D. J.,** Efeito da Temperatura Sobre a Exigência de Proteína e Energia em Dietas para Alevinos de Pacu *Piaractus mesopotamicus* (Holmberg, 1887), Ph.D. thesis, UFSCAR, Sãn Carlos, S.P., 1990.

7. **Carneiro, D. J., Castagnolli, N., Machado, C. I., and Verardtno, M.,** Nutrição de pacu, *Colossoma mitrei* (Berg, 1895) (Pisces, Characidae). I. Níveis de proteína dietária, in *Proc. Simp. Bras. Aqüicul. 3*, UFSCAR, São Carlos, S.P., 1984, 105.

8. **Carneiro, D. J. and Castagnolli, N.,** Nutrição de pacu, *Colossoma mitrei* (Berg, 1895). II. Digestibilidade aparente de proteína em dietas isocalóricas, in *Proc. Simp. Bras. Aqüicul. 3*, UFSCAR, São Carlos, S.P., 1984, 125.

9. **Carneiro, D. J., Castagnolli, N., Machado, C. R., and Verardino, M.,** Nutrição de pacu *Colossoma mitrei* (Berg, 1895). III. Níveis de energia metabolizável em dietas isoprotéicas, in *Proc. Simp. Bras. Aqüicul. 3*, UFSCAR, São Carlos, S.P., 1984, 133.

10. **Cantelmo, O. A. and Sousa, J. A.,** Alimentação do pacu *Colossoma mitrei* em diferentes proporções de proteína animale vegetal, in *Síntese dos Trabalhos Realizados com Espécies do Gênero Colossoma*, CEPTA, Pirassununga, S.P., 1987, 28.

11. **Brenner, M.,** Determinação da Exigência de Proteína do Pacu (*Colossoma mitrei* Berg, 1895), M.S. thesis, Viçosa, U.F.V., 1988.

12. **Cantelmo, O. A. and Sousa, J. A.,** Alimentação em diferentes níveis proteícos para o desenvolvimento inicial do pacu em tanques fertilizados, in *Síntese dos Trabalhos Realizados com Espécies do Gênero Colossoma*, CEPTA, Pirassununga, S.P., 1987, 26.

13. **Carneiro, D. J.,** personal communication, 1990.

14. **Pezzato, L. E.,** Efeito de Diferentes Níveis de Gordura de Origem Animal e Vegetal Sobre o Desempenho, Assimilação, e Deposição de Matéria Graxa Pelo Pacu, *Piaractus mesopotamicus* (Holmberg, 1887), Ph.D. thesis, UNESP, FMVA Jaboticabal, S.P., 1990.

15. **Castagnolli, N. and Donaldson, E. M.,** Induced ovulation and rearing of pacu, *Colossoma mitrei*, *Aquaculture*, 25, 275, 1981.

16. **Pinto, M. L. G. and Castagnolli, N.,** Desenvolvimento inicial do pacu, *Colossoma mitrei* (Berg, 1895), in *Proc. Simp. Bras. Aqüicul. 3*, UFSCAR, São Carlos, S.P., 1984, 523.

17. **Brulé, A. O. and Castagnolli, N.,** Produção de organismos como fonte de alimentos para larvas e alevinos (Rev. Bibliog.), *Vet. e Zoot.*, 1, 7, 1985.

18. **De Stefani, M. V. and Castagnolli, N.,** Produção de *Diaphanosoma* sp. (Cladocera) com diferentes fertilizantes e alimentos, in *Proc. Simp. Latinoam. de Acuicul. 6; Simp. Bras. Aqüicul. 5*, Florianopolis, S.C., 1990, 260.

19. **Yamanaka, N.,** Descrição, Desenvolvimento e Alimentação de Larvas e Pré-juvenis de Pacu (*Piaractus mespotamicus*) (Teleostei, Characidae), Mantidos em Confinamento, Ph.D. thesis, USP. Inst. Bioc., São Paulo, S.P., 1988.

20. **Sipauba Tavares, L. H.,** Utilização de Plâncton na Alimentação de Larvas e Alevinos de Peixes, Ph.D. thesis, UFSCAR, São Carlos, S.P., 1988.

21. **Dias, R. C. R., Castagnolli, N., and Carneiro, D. J.,** Alimentação de larvas de pacu, *Piaractus mesopotamicus* (Holmberg, 1887) com dietas naturais e artificiais, in *Proc. Simp. Latinoam. Acuicul. 6; Simp. Bras. Aqüicul. 5*, Florianopolis, S.C., 1990, 500.

CHANNEL CATFISH, *ICTALURUS PUNCTATUS*

Robert P. Wilson

INTRODUCTION

The channel catfish is the most widely cultured foodfish in the U.S. Channel catfish were originally native to states within the Mississippi Valley and those along the Gulf of Mexico. However, they have been widely introduced throughout the U.S. during the last century. State and federal hatcheries have been providing channel catfish fingerlings to stock public and private waters for sport fisheries for well over 50 years.

Commercial production of channel catfish was developed in the southern U.S. during the late 1950s and early 1960s. The rapid expansion of the industry did not take place until the late 1970s and early 1980s. Current estimates (July 1990) indicate that 63,761 ha of water in 17 states are in production of farm-raised catfish. An additional 5300 ha of production ponds are currently under construction and will be in production next year. Mississippi continues to lead the nation in catfish production, with 37,652 ha or 59% of the total. The other major states include Arkansas with 12.1%, Alabama with 11.4%, and Louisiana with 7.3% of the total. Commercial production of channel catfish is also being developed in Mexico and Italy. Channel catfish are a warmwater species, thus a growing season of at least 6 months or longer with water temperatures above 23°C are needed for their economic culture.

Essentially, all of the farm-raised catfish are produced in static ponds. Initially, pond size varied from 0.4 to 16 ha in size. Subsequent research indicated that the most economical size was about 8 ha. Prior to 1970, fish were stocked in ponds during the spring, raised through a single growing season, and harvested during the fall by draining the pond. The next spring the pond was refilled and a new crop of fish was stocked. The major problem with this type of production system was the seasonal supply of fish and the inability to provide fresh fish throughout the year.

In the early 1970s, the industry developed a different production system that involved multiple harvesting, or topping, throughout the year. This continuous production system involves selectively removing marketable-sized fish, allowing the submarketable-sized fish to remain in the pond, and then an equivalent number of small fingerlings are added to replace the large fish removed. This production system allows the producer to maintain a consistently higher standing crop of fish throughout the year. This more fully utilizes the carrying capacity of the pond and assures a continual supply of marketable fish. These ponds are operated indefinitely without draining. The major problem with this type of production system is the difficulty of maintaining an inventory of the number of fish in the pond. A small percentage of catfish are also produced in densely stocked cages and raceways.

The industry depends on a continual supply of fingerlings or juvenile fish for stocking. Even though certain catfish producers specialize in fingerling production, a large number of catfish farmers maintain broodstock and produce their own fingerlings as part of their commercial operation. Channel catfish normally spawn in late spring and early summer. The broodfish are maintained at low density in ponds provided with spawning containers. The spawning containers are normally checked daily. When an egg mass is found, it is removed and transferred to a hatching trough. Hatching occurs in 5 to 8 d, depending on water temperature. Following hatching, the yolk-sac fry are held in a rearing trough and then the swim-up fry are fed a nutritionally complete high-protein fry feed. Stocking density affects the growth rate of the fry; therefore, the standing crop in the nursery ponds must be reduced at intervals by moving the larger fish to fingerling ponds. Fingerlings are normally stocked in production ponds when they reach 15 to 20 cm in length.

Initially, catfish farmers stocked 2000 to 4000 fingerlings per ha, with annual yields of 1000 to 2000 kg/ha. However, with improved management strategies, high-quality feeds, supplemental aeration, and the development of the continuous production system, stocking rates and production yields have increased dramatically. Current stocking rates range from about 15,000 to 25,000 fish/ha, consisting of a mixture of small, medium, and large fish.

LABORATORY CULTURE

Channel catfish fingerlings are relatively easy to culture in the laboratory. Catfish fry or 1 to 5 g fingerlings are normally moved into a holding tank in the laboratory from a nursery pond. Two weeks prior to the beginning of a nutritional study, fish are moved from the holding tank to the flow-through aquaria that will be used for the nutritional study. The nutrition laboratory should be equipped with an adequate supply of high-quality water, i.e., either well water or dechlorinated municipal water. The water should be supplied to the aquaria at a constant temperature and constant pressure to assure a constant flow rate. Supplemental aeration may or may not be necessary, depending on the water delivery system. Most nutritional studies are conducted at a water temperature of about 27°C. A constant diurnal light/dark cycle of about 14:10 should also be maintained.

The experimental fish should be conditioned or acclimated to the experimental conditions and fed a conditioning diet for about 2 weeks before starting the nutritional study. The conditioning diet should be very similar, if not the same as the basal diet to be used in the study. The diet should be fed at least twice per day at either a fixed percentage of body weight that is near satiation or at satiation. Feeding should be conducted at approximately the same time each day, 7 d/week, so the fish become trained to take the feed immediately after it is dropped into the aquaria. If the fish are to be fed at a fixed percentage of body weight/day, then it is necessary to weigh the fish weekly and to adjust their feed allotment accordingly. If, however, the fish are being fed to satiation, then biweekly weighing is recommended in order to evaluate the progress of the experiment. The length of the nutritional experiment varies with the nutrient being evaluated. Normally, 8 weeks are adequate to obtain statistical differences in growth parameters with macronutrients. However, up to 20 weeks may be necessary when evaluating micronutrients.

Channel catfish readily accept purified diets, such as the example presented in Table 1. In fact, weight gains of fingerlings fed purified diets are usually much higher than those obtained by feeding practical diets under laboratory conditions. Various protein sources, such as vitamin-free casein and gelatin (4:1), hexane-extracted whole chicken-egg powder, egg-white protein, isolated soy protein, and blood fibrin have been used successfully in basal diets to evaluate various nutrient requirements of catfish.

NUTRIENT REQUIREMENTS

PROTEIN AND AMINO ACIDS

Estimations of the protein requirement of channel catfish have been based on total weight gains of fish fed practical and purified diets. These studies have indicated optimum dietary levels of crude protein ranging from 25 to 45%. These differences are probably due to differences in size of the fish, water temperature, natural food available in the ponds, fish stocking density, daily feed allowance, amount of nonprotein energy in the feed, and the quality of the dietary protein.

Most of the early work on determining the protein requirements of catfish was conducted in earthen ponds using practical diets. Those studies were carried out at low fish stocking rates, and the fish were grown to harvestable size. The results of these studies indicated the crude

TABLE 1
Example of a Purified Diet for Channel Catfish[1,2]

Ingredient	Percent dry weight
Casein (vitamin free)	24.40
Gelatin	6.00
Dextrin	33.10
Cellulose	18.00
Corn oil	5.00
Cod liver oil	5.00
Vitamin premix[a]	3.00
Mineral premix[b]	4.00
Calcium carbonate	1.00
Carboxymethyl cellulose	0.50

[a] Contained as g/kg premix: choline chloride, 75; ascorbic acid, 45; inositol, 5; niacin, 4.5; calcium d-pantothenate, 3; riboflavin, 1; pyridoxine·HCl, 1; thiamin·HCl, 1; biotin, 0.02; folic acid, 0.09; Vitamin B_{12}, 0.00135; retinyl acetate, 0.6; cholecalciferol, 0.083; menadione, 1.67; DL-α-tocopheryl acetate (powder 250 U/g), 8; cellulose, 854.04.

[b] Contained as g/kg premix: $Ca(H_2PO_4)_2 \cdot H_2O$, 135.49; $Ca(O_2CCHOHCH_3)_2$, 327; $FeC_6H_5O_7 \cdot H_2O$, 29.7; $MgSO_4 \cdot 7H_2O$, 132; K_2HPO_4, 239.8; $NaH_2PO_4 \cdot H_2O$, 87.2; NaCl, 43.5; $AlCl_3$, 0.15; KI, 0.15; CuCl, 0.2; $MnSO_4 \cdot H_2O$, 0.8; $CoCl_2 \cdot 6H_2O$, 1; $ZnSO_4 \cdot 7H_2O$, 3; Na_2SeO_3, 0.011.

protein requirement to be 25 to 30%.[3-6] Growth and feed conversion of fish stocked at higher stocking rates were found to be improved by increasing the dietary crude protein from 32 to 45% when the feeding rate was restricted to help maintain water quality.[7]

Studies in controlled environments have indicated that the protein requirement ranges from 25 to 36%, depending on fish size. When 114 g or larger fish were fed to satiation, 25% dietary crude protein was adequate for maximum growth. However, when the feeding rate was restricted, higher protein levels were beneficial.[8] Smaller fish required higher protein levels and grew best at 35% dietary crude protein. Maximum growth rates were observed in small (7 g) catfish fed purified diets containing 36% crude protein and 341 kcal/100 g diet.[9] However, maximum protein deposition was observed in fish fed 24% crude protein and 275 kcal/100 g diet. Gatlin et al.[10] have determined the protein requirement for maintenance and maximum growth of small catfish fed purified diets. The protein maintenance requirement was found to be 1.00 to 1.32 g protein/kg body weight/d, and the value for maximum growth was 8.75 g protein/kg body weight/d. This latter value indicates that a 29% crude protein diet should be adequate for catfish when fed at a feeding rate of 3% of body weight/d. Most commercial catfish feeds contain 32% crude protein. However, some fish farms contract for 28, 30, and 36% crude protein feeds.

Channel catfish have been shown to require the same 10 essential amino acids as other fishes.[11] The quantitative requirements for these essential amino acids are summarized in Table 2. These requirement values were obtained by feeding graded levels of each respective amino acid in a test diet containing a mixture of casein, gelatin, and amino acids formulated so that the amino acid profile was identical to whole chicken-egg protein, except for the amino acid being tested. These amino acid test diets had to be neutralized to pH 7.0 for maximum utilization.[14] A linear increase in growth rate was observed with increasing amino acid intake up to a breakpoint, which corresponded to the requirement of that specific amino acid. Most of the requirement values were based solely on the conventional growth response curve. In certain cases, changes in serum amino acid levels were used to confirm the growth response data. A highly significant correlation was found between the essential amino acid requirement

TABLE 2
Quantitative Amino Acid Requirements of
Channel Catfish[a]

Amino acid	Requirement		Ref.
Arginine	4.3	(1.03/24)	12
Histidine	1.5	(0.37/24)	13
Isoleucine	2.6	(0.62/24)	13
Leucine	3.5	(0.84/24)	13
Lysine	5.1	(1.23/24)	14
	5.0	(1.50/30)	15
Methionine	2.3	(0.46/24)	16
Phenylalanine	5.0	(1.20/24)	17
Threonine	2.0	(0.53/24)	18
Tryptophan	0.5	(0.12/24)	18
Valine	3.0	(0.71/24)	13

[a] Requirements are expressed as percent of dietary protein. In parentheses, the numerators are requirements as percent of diet and the denominators are percent total protein in the diet.

pattern and the content of the same amino acids in whole body tissue of the catfish. In addition, no differences were detected in the whole-body amino acid composition of catfish ranging from 30 to 863 g, which was interpreted to indicate that the amino acid requirements when expressed as a percent of dietary protein should not change with increasing size of the fish.[19] No gross deficiency signs were detected in the fish fed the various amino acid-deficient diets, other than a marked reduction in weight gain.

The requirement value for methionine in Table 2 actually represents the total sulfur amino acid requirement, since it was determined in the absence of cystine. Cystine can be formed from dietary methionine, whereas methionine cannot be synthesized from cystine. Thus, the total sulfur amino acid requirement can be met by either methionine alone or a proper mixture of methionine and cystine. Cystine was found to be able to replace or spare 60% of the methionine requirement on a millimole sulfur basis.[16] DL-methionine was found to be utilized as effectively as L-methionine; however, methionine hydroxy analogue was only 26% as effective in promoting growth as L-methionine.[20] Tyrosine was found to be able to replace about 50% of phenylalanine to meet the total aromatic amino acid requirement.[17]

Some adverse interactions may occur when an imbalance of certain structurally similar amino acids are fed to animals. The lysine-arginine antagonism, often observed in other animals, could not be induced in catfish fed either excess lysine in diets adequate or marginal in arginine, or when excess arginine was fed in diets adequate or marginal in lysine.[12] Catfish do appear to be sensitive to dietary imbalances in the branched-chain amino acids leucine, isoleucine, and valine. Feeding diets containing an excess of either leucine or isoleucine in diets deficient in one of the other branched-chain amino acids resulted in depressed weight gain and feed efficiency. This adverse effect could be reversed by supplementation with the deficient amino acid, but not by the addition of other branched chain amino acids.[21] Changes in dietary leucine levels were found to have a marked effect on serum levels of isoleucine and valine. These observations were interpreted to indicate that leucine may facilitate either the tissue uptake or intracellular metabolism of branched-chain amino acids in the catfish.[13]

Early workers were unable to demonstrate the utilization of supplemental amino acids in practical diets for channel catfish.[22,23] Since these earlier studies were conducted before the essential amino acid requirement values were available, it was assumed that the basal diets used may not have been deficient in the amino acid under study. However, only limited success has been reported on the use of supplemental amino acids by channel catfish. For example, supplemental lysine[15] and methionine[24] have been shown to improve weight gains in fish fed peanut meal- and soybean meal-based diets, respectively. Additional research is needed in this important area of amino acid nutrition, because at present, inadequate information is available to recommend the use of amino acid supplementation in commercial catfish feeds.

ENERGY

Actual energy requirements have not been established for the channel catfish. Current recommendations on energy levels used in formulating catfish feeds are based on optimum energy-to-protein ratios. Providing the optimum energy levels in diets for catfish is important because inadequate energy will result in the fish utilizing dietary protein for energy rather than for protein synthesis. Excess energy in the diet may result in decreased nutrient intake by the fish or excessive fat deposition in the fish. Since digestible energy (DE) values were not available when most of the energy studies were conducted, various workers have used estimated DE values based on physiological fuel values.

To date, most studies have indicated a DE requirement of 8 to 9 kcal/g protein to be adequate for fingerling and production size catfish.[9,25-28] These values are based on studies in which fish were fed practical feeds in ponds,[25] fingerling-size catfish were fed purified diets in aquaria,[9] fry- to fingerling-size fish were fed practical diets,[26] various-size catfish ranging from 3 to 266 g were fed to satiation,[27] and fingerling catfish were fed purified diets at three different water temperatures.[28]

Gatlin et al.[10] have determined the energy requirement of fingerling catfish for maintenance and maximum gain. The requirement values were based on weight gain and body composition data from fish fed purified diets at varying daily rations. Energy requirements for maintenance and maximum growth were 15.0 and 17.3 kcal/kg body weight/d, respectively.

CARBOHYDRATES

The nutritional value of various forms of dietary carbohydrate is one of the least understood aspects of fish nutrition. The relative utilization of dietary carbohydrates by fish appears to be associated with the complexity of the carbohydrate. Early studies had indicated that channel catfish utilized higher molecular weight carbohydrates, such as starch or dextrin, more readily than disaccharides or simple sugars.[4,29] These earlier observations were confirmed by Wilson and Poe,[1] who found that corn starch and dextrin served as very good energy sources, whereas glucose, fructose, maltose, and sucrose did not. These data indicated that catfish were unable to utilize dietary mono- or disaccharides as energy sources.

Oral carbohydrate tolerance tests using glucose, maltose, fructose, sucrose, and dextrin were conducted with larger channel catfish. Oral glucose and maltose resulted in a persistent hyperglycemia indicative of a diabetic-like status. Fructose appeared to be poorly absorbed from the intestinal tract and did not appear to be converted to glucose. Oral administration of sucrose was followed by a gradual increase in plasma glucose, with no detectable fructose being absorbed until 6 h after administration. Oral dextrin resulted in less than a twofold increase in plasma glucose, which remained constant from 2 to 4 h after administration and then declined. These observations were considered to be consistent with the hypothesis that certain fishes, including the channel catfish, resemble diabetic animals by having insufficient insulin for maximum carbohydrate utilization.

Catfish have been shown to use higher levels of dietary carbohydrate than certain other fish. Levels of dietary carbohydrate up to 25% have been shown to be utilized as effectively as lipids as an energy source for catfish.[30] Likimani and Wilson[31] observed that channel catfish fed a high-carbohydrate diet exhibited marked stimulation of several lipogenic enzyme activities in both liver and mesenteric adipose tissue. A similar study using coho salmon, *Oncorhynchus kisutch*, failed to demonstrate any stimulatory effect of high dietary carbohydrate on lipogenic enzyme activities in either liver or mesenteric adipose tissue of this fish.[32] Therefore, it appears that species such as the catfish that can readily adapt to high-carbohydrate diets and convert the excess energy into lipids are much more efficient in carbohydrate utilization than those fish lacking that ability.

Although no specific carbohydrate requirement has been established for the catfish, growth rates of fingerling channel catfish were reduced when the fish were fed isonitrogenous, isocaloric semipurified diets containing no dextrin, as compared with fish fed diets containing added dextrin.[30,31] Carbohydrates also serve as the least expensive source of dietary energy and aid in the pelleting quality of practical catfish feeds. Therefore, some form of digestible carbohydrate should be included in catfish diets. A typical commercial catfish feed contains about 25 to 30% carbohydrate.

LIPIDS AND ESSENTIAL FATTY ACIDS

Early workers were unsuccessful in establishing an essential fatty acid (EFA) requirement of channel catfish; however, it had been reported that catfish fed diets containing animal fat or fish oil showed better growth than fish fed diets containing vegetable oils.[33-35] These studies indicated that catfish could effectively utilize beef tallow (high in oleic acid and low in n-3 and n-6 fatty acids) or menhaden oil [moderate in n-3 highly unsaturated fatty acids (n-3 HUFA)], whereas safflower oil (high in linoleic acid) or linseed oil (high in linolenic acid) were poorly utilized. These workers suggested that the poor response to vegetable oils was due to a limited ability to metabolize linoleic acid (18:2n-6) or to competitive inhibition of fatty acid synthesis in the presence of high levels of dietary linolenic acid (18:3n-3). Stickney et al.[36] suggested that the poor performance of catfish fed vegetable oil was due to dietary linolenic acid and not linoleic acid, as first theorized. Satoh et al.[37] reported that supplementation with a mixture of methyl esters of n-3 HUFA or cod liver oil (moderate in n-3 HUFA) to a purified diet enhanced growth of channel catfish. However, catfish fed a diet containing beef tallow as the sole lipid source had a lower growth rate than fish fed a diet with n-3 HUFA. Fatty acid composition of the liver polar lipid fraction indicated that channel catfish could convert 18:3n-3 to docosahexaenoic acid (22:6n-3). Subsequently, Satoh et al.[2] demonstrated that n-3 fatty acids were essential for the channel catfish. Specifically, these workers found the EFA requirement to be met by 1.0 to 2.0% 18:3n-3 or 0.5 to 0.75% n-3 HUFA. Therefore, marine fish oil or linseed oil, which are rich sources of either linolenic acid or n-3 HUFA, should be suitable for supplementation in catfish feeds.

Catfish appear to perform well on a wide range of dietary lipid levels. Studies have been conducted in which catfish were fed diets containing up to 15% or more lipid without conclusive evidence as to which level is best for optimum growth. From a practical viewpoint, lipids should be added to catfish diets as a source of energy. However, there are constraints on the level of dietary lipid that can be used. If the dietary lipid level is too high, undesirable deposits of lipid in edible tissue may occur or deposits of adipose tissue in the visceral cavity may be excessive, which reduces the dressout percentage. A second constraint involves the ability of the feed manufacturer to prepare good-quality feed pellets. If the supplemented fat level exceeds approximately 5%, some difficulty in producing extruded pellets occurs unless the lipid is sprayed on after pelleting. Normally, fat is sprayed on catfish feeds after pelleting to increase dietary energy and to reduce fines. Generally, lipid levels in catfish feeds are about 5 to 6%, approximately 3 to 4% contained in the feed ingredients, with the remaining 1 to 2% being sprayed on the finished pellets.

VITAMINS

The qualitative vitamin requirements of catfish have been determined by feeding purified diets deficient in the vitamin under study and comparing various growth parameters and pathology to fish fed nutritionally complete diets. Quantitative vitamin requirements were

TABLE 3
Vitamin Deficiency Signs and Minimum Levels of Vitamins Required to Prevent Signs of Deficiency in Catfish

Vitamin	Requirement[a] mg/kg	Deficiency signs[b]
Vitamin A	1000–2000 IU	Exophthalmia, edema, acities
Vitamin D	250–500 IU	Low bone ash
Vitamin E	50	Skin depigmentation, exudative diathesis, muscle dystrophy, erythrocyte hemolysis, splenic and pancreatic hemosiderosis
Vitamin K	R	Skin hemorrhage, prolonged clotting time
Thiamine	1	Dark skin color, neurological disorders
Riboflavin	9	Short-body dwarfism
Pyridoxine	3	Greenish-blue coloration, tetany, nervous disorders
Pantothenic acid	15	Clubbed gills, anemia, and eroded skin, lower jaw, fins, and barbels
Niacin	14	Anemia, lesions of skin and fins, exophthalmia
Biotin	R	Anemia, skin depigmentation, reduced liver pyruvate carboxylase activity
Folic acid	R	Anemia
B_{12}	R	Anemia
Choline	400	Fatty liver, hemorrhagic kidneys
Inositol	NR	None
Ascorbic acid	60	Scoliosis, lordosis, reduced bone collagen, internal and external hemorrhage

[a] R = Required but no requirement value determined. NR = Not required.

[b] Anorexia, reduced weight gain, and increased mortality are common vitamin deficiency signs and are not included in the table.

then determined by feeding graded levels of the vitamin under study in purified diets. The dietary level that resulted in normal weight gain with the absence of any deficiency signs was considered to be the minimum dietary requirement for that specific vitamin. These requirement studies were conducted with small fish in aquaria, thus the requirement values were determined under optimal conditions for the fish. Although these values are assumed to be adequate for larger fish, certain vitamin requirements may be affected by other factors, such as size, age, growth rate, and stage of sexual maturity, as well as environmental stressors, such as disease, water temperature, and water quality. Additional research is needed to examine this important area of vitamin nutrition of channel catfish. Dietary requirements and characteristic deficiency signs for each vitamin are summarized in Table 3.

Channel catfish fed a vitamin A-deficient diet for 2 years developed exophthalmia, edema, and hemorrhages in the kidney.[38] The minimum dietary level of vitamin A required for maximal weight gain and to prevent deficiency signs was found to be between 1000 and 2000 IU/kg of diet.[39] Beta-carotene was also shown to be utilized by the catfish as a vitamin A source if present at levels exceeding 2000 IU/kg of diet.[39]

Channel catfish fed a vitamin D-deficient diet for 16 weeks showed reduced weight gain, lower body ash, lower body phosphorus, and lower body calcium compared to controls.[40] In a similar study, Andrews et al.[41] reported that vertebral ash values were not affected by dietary vitamin D levels. Lovell and Li[40] reported that 500 IU of vitamin D/kg of diet was adequate for catfish. However, Andrews et al.[41] reported improved weight gain at 1000 IU of vitamin D/kg of diet. Andrews et al.[41] also demonstrated that vitamin D_3 was utilized more effectively by catfish than vitamin D_2 at dietary levels of 2000 IU/kg of diet and that high levels of vitamin D_3 (20,000 to 50,000 IU/kg of diet) reduced weight gain.

Brown[42] reevaluated the utilization of vitamin D_2 and D_3, and the dietary requirement for D_3, of catfish reared in calcium-free water. Vitamin D_2 was utilized as well as vitamin D_3 up to 1500 IU/kg of diet, but higher levels of vitamin D_2 depressed weight gain and feed efficiency. A dietary level of 250 IU of vitamin D_3/kg of diet resulted in maximum weight gain and feed efficiency. However, dietary levels of 1000 to 2000 IU of vitamin D_3/kg of diet were required to produce maximum levels of serum 25-hydroxyvitamin D concentrations.

Channel catfish fed an α-tocopherol-deficient diet containing oxidized menhaden oil showed reduced growth, muscular dystrophy, fatty livers, anemia, exudative diathesis, and depigmentation in 16 weeks.[43] Fish fed a deficient diet without oxidized menhaden oil showed exudative diathesis, depigmentation, and mortality. Addition of 25 mg α-tocopherol plus 125 mg ethoxyquin or 100 mg α-tocopherol/kg of diet prevented those changes, even in the presence of oxidized menhaden oil. Based on these observations, a vitamin E requirement value of 30 mg/kg of diet was recommended. Subsequent studies by Lovell et al.[44] showed that the above-indicated deficiency signs could be demonstrated in catfish fed diets containing low levels of polyunsaturated fatty acids. These workers also suggested that the above-indicated requirement level of 30 mg/kg of diet was inadequate; however, a detailed requirement study was not conducted.

Wilson et al.[45] did not observe the typical gross vitamin E deficiency signs in catfish fed diets adequate in selenium but deficient in vitamin E during a 20-week study, except for increased erythrocyte hemolysis. Subclinical signs, such as marked multifocal splenic hemosiderosis and mild multifocal hemosiderosis of pancreatic tissue, were observed. A vitamin E requirement of 50 mg DL-α-tocopheryl acetate/kg of diet was determined based on liver microsomal ascorbic acid-stimulated lipid peroxidation data. Gatlin et al.[46] subsequently demonstrated the nutritional interaction between vitamin E and selenium in catfish. These workers observed only subclinical signs of vitamin E deficiency in fish fed vitamin E-deficient diets that were adequate in selenium. Similarly, fish fed selenium-deficient diets that were adequate in vitamin E only developed subclinical signs of selenium deficiency. However, when fish were fed diets deficient in both vitamin E and selenium, classical signs of both vitamin E and selenium deficiency were observed.

No vitamin K requirement has been established for channel catfish. Catfish fed a vitamin-K deficient diet grew normally but did have some hemorrhages in the skin that were attributed to vitamin K deficiency.[38] However, Murai and Andrews[47] were unable to demonstrate a requirement for vitamin K in catfish fed a deficient diet for 30 weeks. The addition of dicumerol to the diet of catfish that had been fed a vitamin K-deficient diet for 24 weeks did not alter prothrombin times.

Dupree[38] observed neurological disorders in channel catfish fed a thiamine-deficient diet for 20 weeks. Murai and Andrews[48] observed anorexia, poor growth, dark coloration of the skin, and increased mortality in catfish fed thiamine-deficient diets for 6 to 8 weeks. These workers determined a requirement value of 1 mg thiamine/kg of diet.

Channel catfish fed riboflavin-deficient diets for 8 weeks developed deficiency signs, including anorexia, poor growth, short-body dwarfism,[49] and cataracts.[38] A dietary level of 9 mg riboflavin/kg of diet is required for maximum growth and prevention of deficiency signs by the catfish.[49]

Channel catfish fed pyridoxine-deficient diets for 6 to 8 weeks exhibited signs of deficiency, including anorexia, tetany, nervous disorders, and a greenish-blue coloration.[50] The dietary level required for maximal growth and prevention of deficiency signs was 3 mg pyridoxine/kg of diet.

Pantothenic acid-deficient catfish exhibited severe anorexia, poor growth, clubbed gills, anemia, high mortality, and eroded skin, lower jaw, fins, and barbels.[51] These workers recommended a level of 10 mg pantothenic acid/kg of diet for maximal growth and prevention of gross deficiency signs. However, the above level proved to be inadequate in commercial catfish feeds, as evidenced by an outbreak of severe gill tissue proliferation indicative of pantothenic acid deficiency in farm-raised catfish in the early 1980s. Wilson et al.[52] reevaluated the requirement and found it to be 15 mg calcium *d*-pantothenate/kg of diet. These workers also observed a high correlation between disease susceptibility and dietary calcium *d*-pantothenate and recommended a level of 30 mg calcium *d*-pantothenate/kg of diet to be used in formulating catfish feeds.

Channel catfish fed a niacin-deficient diet exhibited poor growth, anemia, skin and fin lesions and hemorrhages, exophthalmia, and total mortality in 20 weeks.[53] These workers found that 14 mg niacin/kg of diet was required for maximal growth. Mortality and gross deficiency signs were prevented by 11.6 mg niacin/kg of diet and the anemia was prevented by only 6.6 mg niacin/kg of diet.

Initial efforts were unable to demonstrate a dietary need for biotin by channel catfish fed purified diets at 24°C for 30 weeks.[38] However, biotin deficiency was subsequently induced in catfish fed purified diets containing raw egg white as a source of avidin.[54] These workers observed light skin pigmentation and anemia after 14 weeks and reduced liver pyruvate carboxylase activity after 22 weeks on the biotin-deficient diet containing avidin. Lovell and Buston[55] were able to induce biotin deficiency in catfish fed purified diets, with or without egg white, in 17 weeks. Deficiency signs included anorexia, reduced growth rate, lighter skin color, hypersensitivity, and reduced liver pyruvate carboxylase activity. These deficiency signs were not detected when catfish were fed practical diets containing 0.33 to 0.37 mg biotin/ kg of diet. The ratio of biotin to indigestible dry matter in feces was lower than that in the practical diet, indicating little to no synthesis of biotin by the intestinal microflora in channel catfish. These workers concluded that practical diets for channel catfish made from the commonly used ingredients, such as soybean meal, corn meal, and menhaden fish meal, do not need supplemental biotin.

Dupree[38] fed channel catfish folic acid-deficient diets for 30 weeks and observed reduced growth and increased mortality, but no specific deficiency signs. Since folic acid has been shown to be synthesized by the intestinal bacteria of common carp,[56] it has been suggested that this source may also apply to the catfish.[57] However, a recent study has shown reduced growth, anemia, and increased sensitivity to bacterial infection in catfish fed folic acid-deficient diets.[58]

Channel catfish fed a vitamin B_{12}-deficient diet for 36 weeks exhibited reduced growth but no deficiency signs.[38] Intestinal microbial synthesis of vitamin B_{12} has been demonstrated in channel catfish when dietary cobalt was present.[59] However, these workers did observe reduced hematocrits in catfish fed diets deficient in both vitamin B_{12} and cobalt after 24 weeks. A dietary requirement level for vitamin B_{12} has not been established.

Channel catfish fed choline-deficient diets for 36 weeks had reduced weight gain, enlarged livers, and hemorrhagic areas in the kidney.[38] Catfish fed casein-gelatin diets without choline but containing excess methionine did not develop signs of choline deficiency during a 12-week study.[60] However, when the fish were fed diets containing isolated soybean protein that were just adequate in methionine, a choline deficiency was induced. An optimal level of 400 mg choline/kg of diet based on liver lipid content was determined for catfish under the dietary conditions used. These workers stated that this value would appear to represent a maximum requirement level because it was determined at the lowest level of dietary methionine that could be used while still meeting the total sulfur amino acid requirement of the catfish. These workers also demonstrated that catfish can utilize dietary methionine to spare, in part, their need for dietary choline.

No signs of inositol deficiency were observed when channel catfish were fed purified diets without supplemental myo-inositol and containing an antibiotic to suppress intestinal bacteria synthesis of inositol.[61] This study also showed that *de novo* synthesis of myo-inositol by fingerling channel catfish was adequate for normal growth and maintenance of tissue levels of myo-inositol and to prevent overt signs of myo-inositol deficiency when the vitamin was not included in the diet. Thus, inositol is not considered to be essential for the catfish.

Channel catfish fed ascorbic acid-deficient diets develop the following characteristic deficiency signs: reduced weight gain, scoliosis and lordosis, increased susceptibility to bacterial infections, internal and external hemorrhage, dark skin color, fin erosion, and reduced formation of bone collagen.[62-66] A dietary level of 60 mg ascorbic acid/kg of diet has been shown to be required for normal growth and prevention of deficiency signs;[63,67] however, higher levels have been reported to increase resistance to bacterial infection.[63,68] Exposure of catfish to certain pesticides, such as toxaphene, has been shown to increase the dietary requirement for ascorbic acid.[69]

The role that ascorbic acid plays in disease resistance and/or the immune system of catfish remains unclear. Durve and Lovell[68] demonstrated that resistance of channel catfish to infection from *Edwardsiella tarda* was enhanced when the dietary level of ascorbic acid was increased to five times the requirement for normal growth. Li and Lovell[67] reported that the mortality of catfish fingerlings fed diets containing from 0 to 3000 mg ascorbic acid/kg of diet for 13 weeks and then infected with *Edwardsiella ictaluri* ranged from 0 in fish fed the highest level of ascorbic acid to 100% in fish fed ascorbic acid-deficient diets. There were no differences in antibody production, complement activity, or phagocytic activities among fish fed diets containing 30 to 300 mg ascorbic acid/kg of diet. However, the 3000 mg ascorbic acid/kg dietary level significantly enhanced antibody production and complement activity. Some fish farmers have fed commercial feeds containing megadoses of ascorbic acid in the early spring when the fish are most susceptible to diseases, but the effectiveness of this practice is difficult to measure. Unpublished data from studies with large catfish reared in cages indicate that megadoses of ascorbic acid may not be as beneficial as first suggested.[57]

A major consideration when formulating feeds for fish is that ascorbic acid is very labile and thus readily destroyed in the manufacturing process, especially in extruded feeds. An ethylcellulose-coated ascorbic acid, which improves stability of the vitamin, is therefore typically used to increase retention of the vitamin in fish feeds. Nevertheless, approximately 50% of the supplemental ascorbic acid is destroyed during the manufacture of extruded catfish feeds,[65] and excess ascorbic acid is routinely added to commercial formulations to ensure that an adequate level of the vitamin is retained in the final product.

Various derivatives of ascorbic acid, which are more stable than the parent compound, have been shown to have antiscorbutic activity in catfish. These include L-ascorbate-2-sulfate,[70-72] L-ascorbyl-2-monophosphate,[70] and L-ascorbyl-2-polyphosphate.[72] However, L-ascorbate-2-sulfate does not appear to be utilized as well as certain of the other more stable forms of ascorbic acid by catfish.[71] At the current time it is still more economical to overfortify with the ethylcellulose-coated ascorbic acid than to use any of the more stable forms.

Current recommended levels of vitamins to be used by commercial feed manufacturers in nutritionally complete, extrusion-processed feeds for channel catfish are presented in Table 4.

MINERALS

Channel catfish probably require the same minerals as other animals. In addition, fish utilize inorganic elements to maintain their osmotic balance between body fluids and the water. Some minerals in the water appear to be absorbed by the fish. For example, catfish can usually absorb sufficient calcium from the water to meet their requirement. However, calcium deficiency has been induced when catfish were reared in calcium-free water.[73] The deficiency was characterized by a reduction in growth rate and in bone ash and a dietary calcium requirement of 0.45% was determined under these unusual conditions.

The dietary requirements of several minerals have been determined for the channel catfish, and the results are summarized in Table 5. These requirement values were determined by feeding purified diets limiting in the mineral under study and evaluating various growth parameters and either tissue mineral levels or enzyme activities.

TABLE 4
Recommended Amounts of Vitamins to be Added in Extrusion Processed Feeds for Channel Catfish[a]

Vitamin	Amount per kg (mg)	Amount per ton (mg)
Vitamin A	4,400 IU	4,000,000 IU
Vitamin D$_3$	2,200 IU	2,000,000 IU
Vitamin E	66 IU	60,000 IU
Vitamin K	4.4	4,000
Thiamine	11	10,000
Riboflavin	13.2	12,000
Pyridoxine	11	10,000
Pantothenic acid	35.2	32,000
Nicotinic acid	88	80,000
Folic acid	2.2	2,000
Vitamin B$_{12}$	0.01	9
Choline chloride (70%)	275	250,000
Ascorbic acid	375.6	340,000

[a] Adapted from Robinson, E. H., *Rev. Aquatic Sci.*, 1, 365, 1989. With permission.

TABLE 5
Mineral Deficiency Signs and Minimum Levels of Minerals Required to Prevent Deficiency Signs in Catfish

Mineral	Requirement	Deficiency signs[a]
Calcium	BLD[b]	
	0.45%[c]	Reduced bone ash
Phosphorus	0.45%[d]	Reduced bone mineralization
Magnesium	0.04%	Muscle flaccidity, sluggishness, reduced bone, serum, and whole body magnesium
Sodium	BLD	
Potassium	BLD	
Chloride	BLD	
Zinc	20.00 mg/kg	Reduced serum zinc and serum alkaline phosphatase activity, reduced bone zinc and calcium
Selenium	0.25 mg/kg	Reduced liver and plasma selenium-dependent glutathione peroxidase activity
Manganese	≤2.40 mg/kg	None detected
Iron	30.00 mg/kg	Reduced hemoglobin, hematocrit, erythrocyte count, reduced serum iron and transferrin saturation levels
Copper	5.00 mg/kg	Reduced hepatic copper-zinc superoxide dismutase, reduced heart cytochrome C oxidase activity

[a] Anorexia, poor weight gain, and increased mortality are common deficiency signs for most mineral deficiencies.
[b] Below level of detection under normal rearing conditions.
[c] Determined in calcium-free rearing water.
[d] Expressed as available phosphorus.

Channel catfish fed phosphorus-deficient diets exhibit poor weight gain, feed conversion, and bone mineralization.[74-76] The requirement expressed as available phosphorus has been shown to be about 0.45% of the diet.[75,76] Since catfish appear to regulate their calcium status via their gills, the concept of an optimum dietary calcium-to-phosphorus ratio does not appear to be important. Deficiency signs of magnesium in catfish include reduced weight gain, anorexia, sluggishness, muscle flaccidity, high mortality, and depressed levels of magnesium in bone, serum, and whole body tissue.[77] A dietary level of 0.04% magnesium was determined to support normal growth and to prevent the deficiency signs. Efforts to induce a deficiency in sodium, potassium, and chloride have been unsuccessful.[78] No effect on either growth rates or whole-body tissue levels of sodium, potassium, or chloride could be detected in catfish fed diets supplemented or unsupplemented with these elements.

Channel catfish fed zinc-deficient diets had reduced weight gain, reduced appetite, serum zinc levels, serum alkaline phosphatase activity, and bone zinc and calcium levels.[79] A dietary requirement value of 20 mg zinc/kg of diet was determined with purified diets. Since it has been well established that phytic acid reduces the availability of zinc in other animals, the requirement was reassisted using a practical diet containing approximately 1.1% phytic acid from soybean meal and rice bran.[80] A supplemental level of 150 mg zinc/kg of diet was needed in the practical diet containing phytic acid in order to met the dietary zinc requirement of the fish.

Feeding diets devoid in selenium to catfish resulted in reduced weight gain and lowered liver and plasma selenium-dependent glutathione peroxidase activity.[81] A requirement value of 0.25 mg selenium/kg of diet was determined for catfish fed vitamin E-adequate diets. Gatlin and Wilson[82] were unsuccessful in inducing a manganese deficiency in fingerling channel catfish. The manganese level contained in the basal diet of 2.4 mg manganese/kg of diet appeared to be adequate for normal growth. Catfish fed iron-deficient diets had decreased weight gain, feed efficiency, hemoglobin, hematocrit, erythrocyte count, and serum iron and transferrin saturation values.[83] A dietary iron requirement of 30 mg/kg of diet was determined for catfish. Copper deficiency in catfish resulted in decreased hepatic copper-zinc superoxide dismutase and heart cytochrome C oxidase activities.[84] A requirement value of 5 mg copper/kg of diet was determined.

No requirement information is available on either cobalt or iodine for the channel catfish. Catfish have been shown to utilize dietary cobalt for vitamin B_{12} synthesis by intestinal microorganism;[59] however, no requirement study has been conducted. The iodine requirement of salmonids has been estimated to be from 1 to 5 mg iodide/kg of diet.[85] It is assumed that the iodine requirement for catfish lies with this range.

Mineral supplementation is added to practical catfish feeds. Natural feedstuffs are considered to be adequate in magnesium, sodium, potassium, and chloride. Calcium and phosphorus are normally added as dicalcium phosphate, mainly as a source of available phosphorus. Selenium is normally supplemented in the vitamin premix at a level to provide 0.1 mg/kg of feed. A trace mineral premix should be used that provides the following minerals in the feed (mg/kg): manganese, 25; iodine, 2.4; copper, 5; zinc, 200; iron, 30; and cobalt, 0.05.

DIGESTIBILITY AND NUTRIENT AVAILABILITY

The nutritional value of a diet or feed is determined ultimately by the ability of the fish to digest and utilize it. Information on nutrient digestibility has lagged behind the many advances made in channel catfish nutrition during the last few years. Only a limited number of investigators have attempted to conduct these types of studies because of many technical difficulties. Determining nutrient digestibility involves measuring the amount of a specific nutrient ingested by the fish and then subtracting that which is present in the feces following

TABLE 6
Average Percent Digestibility Coefficients for Feedstuffs
Commonly Used in Catfish Feeds[91]

Ingredient	International feed number	Crude protein	Gross energy
Corn, grain	4-02-935	97	57
Cottonseed meal	5-01-621	83	80
Meat meal with bone	5-00-388	61	76
Fish, menhaden meal	5-02-009	85	92
Peanut meal	5-03-650	74	76
Rice bran	4-03-928	73	50
Soybean meal (48%)	5-04-612	97	72
Wheat	4-05-268	92	63

digestion. These types of studies are very difficult with fish, because fecal and other metabolic excretions are normally voided into large quantities of water.

Since total fecal collection from catfish is difficult, digestibility trials have been conducted using the indirect method, which involves feeding diets containing chromic oxide and collecting a representative fecal sample. A change in the relative proportion of a nutrient to the chromic oxide in the feed and feces gives an index of the digestibility of the nutrient. Various methods have been used to collect fecal material for analysis during the digestibility trials.[86-91] Digestible protein and energy coefficients for the feed ingredients typically used in channel catfish feed are presented in Table 6. Most of the values presented are similar to those obtained by other investigators. Amino acid availability data are also available for most feed ingredients that are typically used in catfish feed (Table 7).[92]

Phosphorus availability data are summarized in Table 8. In general, phosphorus is about 40% available from animal sources but only 25 to 30% available from plant sources to the catfish. Most of the phosphorus from plant sources is in the form of phytate phosphorus, which limits its availability. Inorganic phosphates are generally highly available to the catfish.

PRACTICAL DIET FORMULATION

Adequate nutritional information is available to formulate high-quality practical feeds for channel catfish. Catfish feeds were formerly classed as supplemental or complete. Supplemental feeds were designed for low stocking density ponds and contained only the macronutrients, i.e., protein and energy, with the assumption that micronutrient requirements would be met from consumption of natural food. A complete feed is one that is formulated to supply adequate amounts of all nutrients that have been shown or suspected to be essential for the catfish. Present stocking densities of catfish are such that the nutrient contribution of natural food in the culture system is minimal and, essentially, all commercial catfish feeds are formulated to be complete.

Most catfish feeds are produced as an extruded or floating feed. Extruded or floating feeds are preferred by the catfish farmer primarily as a management tool, because their use allows the culturist to observe the feeding behavior of his fish. The extrusion process also improves the starch digestibility of cereal grains, reduces fines in the final product, and improves the water stability of the pellets. An example of a typical commercial catfish feed formulation is presented in Table 9. Various other feedstuffs, such as meat and bone meal, peanut meal, cottonseed meal, and milo, have also been used in catfish feeds.

In general, only a limited number of feed types are used by the catfish industry. A fry feed is normally prepared to contain from 40 to 50% crude protein, with about 50% of the protein coming from fish meal. Fry feeds are usually formulated and then ground to a fine mash

TABLE 7
**Average Percent True Amino Acid Availability Values
for Common Feedstuffs Used in Catfish Feeds[92]**

Amino acid	Menhaden fish meal	Meat meal with bone	Soybean meal	Corn	Cottonseed meal	Rice bran	Peanut meal	Wheat middlings
Alanine	89.0	72.5	81.7	83.9	73.3	88.1	92.5	89.7
Arginine	91.0	87.9	96.8	82.0	90.6	94.2	97.7	95.1
Aspartic acid	76.4	61.5	82.0	69.3	82.2	91.0	90.9	90.5
Glutamic acid	84.8	75.1	83.9	86.7	85.4	93.1	92.1	94.9
Glycine	85.1	71.8	75.7	66.3	76.9	87.7	81.3	90.4
Histidine	84.5	82.2	87.9	90.3	81.6	83.4	89.4	94.5
Isoleucine	87.1	80.8	79.7	67.9	71.7	87.5	93.3	87.8
Leucine	89.0	82.4	83.5	87.5	76.4	90.5	95.1	89.9
Lysine	86.4	86.7	94.1	96.5	71.2	94.7	94.1	96.3
Methionine	83.1	80.4	84.6	70.5	75.8	88.2	91.2	82.8
Phenylalanine	87.3	85.4	84.2	81.8	83.5	89.5	96.0	93.0
Proline	83.8	77.9	79.9	83.9	76.2	87.1	91.4	91.8
Serine	86.6	69.8	89.6	77.6	82.5	94.7	92.7	92.1
Threonine	87.4	76.3	82.2	69.8	76.7	88.2	93.4	89.1
Tyrosine	88.8	83.1	83.3	77.5	73.4	93.7	95.5	89.1
Valine	87.1	80.8	78.5	74.4	76.1	89.2	93.3	90.1
Average	86.1	78.4	84.2	79.1	78.3	90.1	92.4	91.1

TABLE 8
Average Percent Phosphorus Availability from
Various Sources for Channel Catfish

Source	International feed number	Availability	Ref.
Phosphates			
Sodium phosphate			
Mono basic	6-04-288	90	75
Calcium phosphate			
Mono basic	6-01-082	94	75
Dibasic	6-01-080	65	75
Fish meals			
Anchovy	5-01-985	40	75
Menhaden	5-02-009	39	75
Protein sources			
Egg albumin		71	76
Casein	5-01-162	90	76
Plant sources			
Wheat middlings	4-05-205	28	75
Corn, grain	4-26-023	25	75
Soybean meal, 44%	5-20-637	50	75
Soybean meal, 48%	5-04-612	29,54	75,76

TABLE 9
Composition of a Typical 32% Crude Protein,
Extruded Catfish Feed[57]

Ingredient	International feed number	Percent
Menhaden fish meal	5-02-009	8.0
Soybean meal (48%)	5-04-612	48.2
Corn	4-02-935	31.2
Rice bran or wheat middlings	4-05-205	10.0
Dicalcium phosphate	6-01-080	1.0
Fat (sprayed on finished feed)		1.5
Trace mineral mix[a]		0.05
Vitamin mix[b]		0.05
Ascorbic acid[c]		0.038

[a] See mineral section for recommended composition.
[b] See Table 4 for recommended composition.
[c] Ethylcellulose coated, contains 98% ascorbic acid.

without being pelleted. These meals should be overfortified with vitamins and coated with oils or fats to enhance their floatation and to retard nutrient leaching. Fingerling feeds are normally produced by crumbling and screening production-type feeds. Production feeds, such as the example presented in Table 9, are used to feed fingerlings up to marketable-size fish. Generally, broodstock fish are also fed the same feed as production-size fish. Feed consumption by catfish is affected both by fish size and water temperature. These effects are evident in the recommended feeding rates presented in Table 10.

TABLE 10
Suggested Maximum Feeding Rates and Feeding Frequencies for Fry or Small Fingerling and for Food-Size Channel Catfish at Different Water Temperatures[93]

Water temperature (°C)	Fry or fingerlings		Food-size fish	
	Frequency	Rate (%)	Frequency	Rate (%)
31° and above	2 times/day	2	1 time/day	1
26–30°	4 times/day	6	2 times/day	3
20–25°	2 times/day	3	1 time/day	2
14–19°	1 time/day	2	1 time/day	2
10–13°	Alternate days	2	Alternate days	1
9° and below	3rd to 4th day	1	3rd to 4th day	0.5

REFERENCES

1. **Wilson, R. P. and Poe, W. E.,** Apparent inability of channel catfish to utilize dietary mono- and disaccharides as energy sources, *J. Nutr.*, 117, 280, 1987.
2. **Satoh, S., Poe, W. E., and Wilson, R. P.,** Effect of dietary n-3 fatty acids on weight gain and liver polar lipid fatty acid composition of fingerling channel catfish, J. *Nutr.*, 119, 23, 1989.
3. **Tiemeier, O. W., Deyoe, C. W., and Wearden, S.,** Effects on growth of fingerling channel catfish of diets containing two energy and two protein levels, *Trans. Kansas Acad. Sci.*, 68, 180, 1965.
4. **Simco, R. A. and Cross, F. B.,** Factors affecting the growth and production of channel catfish, *Ictalurus punctatus, U. Kansas Mus. Nat. Hist.*, 17, 191, 1966.
5. **Deyoe, C. W., Tiemeier, O. W., and Suppes, C.,** Effects of protein, amino acid levels and feeding methods on growth of fingerling channel catfish, *Prog. Fish-Cult.*, 30, 187, 1968.
6. **Hastings, W. H. and Dupree, H. K.,** Formula feeds for channel catfish, *Prog. Fish-Cult.*, 31, 187, 1969.
7. **Lovell, R. T.,** Fish feed and nutrition: how much protein in feeds for channel catfish, *The Commercial Fish Farmer*, March-April, 40, 1975.
8. **Page, J. W. and Andrews, J. W.,** Interactions of dietary levels of protein and energy on channel catfish, *J. Nutr.*, 103, 1339, 1973.
9. **Garling, D. L., Jr. and Wilson, R. P.,** Optimum dietary protein to energy ratio for channel catfish fingerlings, *Ictalurus punctatus, J. Nutr.*, 106, 1368, 1976.
10. **Gatlin, D. M., III, Poe, W. E., and Wilson, R. P.,** Protein and energy requirements of fingerling channel catfish for maintenance and maximum growth, *J. Nutr.*, 116, 2121, 1986.
11. **Dupree, H. K. and Halver, J. E.,** Amino acids essential for the growth of channel catfish, *Ictalurus punctatus, Trans. Am. Fish. Soc.*, 99, 90, 1970.
12. **Robinson, E. H., Wilson, R. P., and Poe, W. E.,** Arginine requirement and apparent absence of a lysine-arginine antagonist in fingerling channel catfish, *J. Nutr.*, 111, 46, 1981.
13. **Wilson, R. P., Poe, W. E., and Robinson, E. H.,** Leucine, isoleucine, valine and histidine requirements of fingerling channel catfish, *J. Nutr.*, 107, 166, 1980.
14. **Wilson, R. P., Harding, D. E., and Garling, D. L., Jr.,** Effect of dietary pH on amino acid utilization and the lysine requirement of fingerling channel catfish, *J. Nutr.*, 107, 166, 1977.
15. **Robinson, E. H., Wilson, R. P., and Poe, W. E.,** Re-evaluation of the lysine requirement and lysine utilization of fingerling channel catfish, *J. Nutr.*, 110, 2313, 1980.
16. **Harding, D. E., Allen, O. W., Jr., and Wilson, R. P.,** Sulfur amino acid requirement of channel catfish: L-methionine and L-cystine, *J. Nutr.*, 107, 2031, 1977.
17. **Robinson, E. H., Wilson, R. P., and Poe, W. E.,** Total aromatic amino acid requirement, phenylalanine requirement and tyrosine replacement value for fingerling channel catfish, *J. Nutr.*, 110, 1805, 1980.
18. **Wilson, R. P., Allen, O. W., Jr., Robinson, E. H., and Poe, W. E.,** Tryptophan and threonine requirements of fingerling channel catfish, *J. Nutr.*, 108, 1595, 1978.
19. **Wilson, R. P. and Poe, W. E.,** Relationship of whole body and egg essential amino acid patterns to amino acid requirement patterns in channel catfish, *Ictalurus punctatus, Comp. Biochem. Physiol.*, 80B, 385, 1985.

20. **Robinson, E. H., Allen, O. W., Jr., Poe, W. E., and Wilson, R. P.,** Utilization of dietary sulfur compounds by fingerling channel catfish: L-methionine, DL-methionine, methionine hydroxy analogue, taurine and inorganic sulfate, *J. Nutr.*, 108, 1932, 1978.
21. **Robinson, E. H., Poe, W. E., and Wilson, R. P.,** Effects of feeding diets containing an imbalance of branched-chain amino acids on fingerling channel catfish, *Aquaculture*, 37, 51, 1984.
22. **Andrews, J. W. and Page, J. W.,** Growth factors in the fish meal component of catfish diets, *J. Nutr.*, 104, 1091, 1974.
23. **Andrews, J. W., Page, J. W., and Murray, M. W.,** Supplementation of a semipurified casein diet for catfish with free amino acids and gelatin, *J. Nutr.*, 107, 1153, 1977.
24. **Murai, T., Ogata, H., and Nose, T.,** Methionine coated with various materials supplemented to soybean meal diet for fingerling carp *Cyprinus carpio* and channel catfish *Ictalurus punctatus*, *Bull. Jpn. Soc. Sci. Fish.*, 48, 85, 1982.
25. **Lovell, R. R. and Prather, E. E.,** Response of intensively fed catfish to diets containing various protein to energy ratios, *Proc. S.E. Assoc. Fish Wildl. Agen.*, 27, 455, 1973.
26. **Winfree, R. A. and Stickney, R. R.,** Starter diets for channel catfish: effects of dietary protein on growth and carcass composition, *Prog. Fish-Cult.*, 46, 79, 1984.
27. **Mangalik, A.,** Dietary Energy Requirements for Channel Catfish, Ph.D. dissertation, Auburn University, Auburn, AL, 1986.
28. **Masser, M. P.,** Effects of Temperature and Dietary Energy/Protein Ratio on Growth of Channel Catfish, Ph.D. dissertation, Texas A&M University, College Station, 1986.
29. **Dupree, H. K.,** Carbohydrate molecular size, in *Progress in Sport Fisheries Research, 1965*, U.S. Bureau of Sport Fisheries and Wildlife, Washington, D.C., 38, 129, 1966.
30. **Garling, D. L., Jr. and Wilson, R. P.,** Effects of dietary carbohydrate-to-lipid ratios on growth and body composition of fingerling channel catfish, *Prog. Fish-Cult.*, 39, 43, 1977.
31. **Likimani, T. A. and Wilson, R. P.,** Effects of diet on lipogenic enzyme activities in channel catfish hepatic and adipose tissue, *J. Nutr.*, 112, 112, 1982.
32. **Lin, H., Romsos, D. R., Tack, P. I., and Leveille, G. A.,** Influence of dietary lipid on lipogenic enzyme activities in coho salmon, *Oncorhynchus kisutch* (Walbaum), *J. Nutr.*, 107, 946, 1977.
33. **Stickney, R. R. and Andrews, J. W.,** Combined effects of dietary lipids and environmental temperature on growth, metabolism and body composition of channel catfish (*Ictalurus punctatus*), *J. Nutr.*, 101, 1703, 1971.
34. **Stickney, R. R. and Andrews, J. W.,** Effects of dietary lipids on growth, food conversion, lipid and fatty acid composition of channel catfish, *J. Nutr.*, 102, 249, 1972.
35. **Yingst, W. L. and Stickney, R. R.,** Effects of dietary lipids on fatty acid composition of channel catfish fry, *Trans. Am. Fish. Soc.*, 111, 90, 1079.
36. **Stickney, R. R., McGeachin, R. B., Lewis, D. H., and Marks, J.,** Response of young channel catfish to diets containing purified fatty acids, *Trans. Am. Fish. Soc.*, 112, 665, 1983.
37. **Satoh, S., Poe, W. E., and Wilson, R. P.,** Studies on the essential fatty acid requirement of channel catfish, *Ictalurus punctatus*, *Aquaculture*, 79, 121,1989.
38. **Durpee, H. K.,** Vitamins essential for growth of channel catfish, *Ictalurus punctatus*, U.S. Bureau of Sport Fish. and Wildlife, Tech. Paper 7, 1966
39. **Durpee, H. K.,** Dietary requirement of vitamin A acetate and beta carotene, in *Progress in Sport Fishery Research, 1969*, Resource Pub. No. 88, Bureau of Sport Fish. and Wildlife, Washington, D.C., 1970.
40. **Lovell, R. T. and Li, Y.-P.,** Essentiality of vitamin D in diets of channel catfish (*Ictalurus punctatus*), *Trans. Am. Fish. Soc.*, 107, 809, 1978.
41. **Andrews, J. W. and Murai, T.,** Effects of dietary cholecalciferol and ergocalciferol on catfish, *Aquaculture*, 19, 49, 1980.
42. **Brown, P. B.,** Vitamin D Requirement of Juvenile Channel Catfish Reared in Calcium-Free Water, Ph.D. dissertation, Texas A & M University, College Station, 1988.
43. **Murai, T. and Andrews, J. W.,** Interactions of dietary α-tocopherol, oxidized menhaden oil and ethoxyquin on channel catfish (*Ictalurus punctatus*), *J. Nutr.*, 104, 1416, 1974.
44. **Lovell, R. R., Miyazaki, T., and Rabegnator, S.,** Requirement for α-tocopherol by channel catfish fed diets low in polyunsaturated triglycerides, *J. Nutr.*, 114, 894, 1984.
45. **Wilson, R. P., Bowser, P. R., and Poe, W. E.,** Dietary vitamin E requirement of fingerling channel catfish, *J. Nutr.*, 114, 2053, 1984.
46. **Gatlin, D. M., III, Poe, W. E., and Wilson, R. P.,** Effects of singular and combined dietary deficiencies of selenium and vitamin E on fingerling channel catfish (*Ictalurus punctatus*), *J. Nutr.*, 116, 1061, 1986.
47. **Murai, T. and Andrews, J. W.,** Vitamin K and anticoagulant relationships in catfish diets, *Bull. Jpn. Soc. Sci. Fish.*, 43, 785, 1977.
48. **Murai, T. and Andrews, J. W.,** Thiamin requirement of channel catfish fingerlings, *J. Nutr.*, 108, 176, 1978.
49. **Murai, T. and Andrews, J. W.,** Riboflavin requirement of channel catfish fingerlings, *J. Nutr.*, 108, 1512, 1978.

50. **Andrews, J. W. and Murai, T.,** Pyridoxine requirements of channel catfish, *J. Nutr.*, 109, 533, 1979.
51. **Murai, T. and Andrews, J. W.,** Pantothenic acid requirement of channel catfish fingerlings, *J. Nutr.*, 109, 1140, 1979.
52. **Wilson, R. P., Bowser, P. R., and Poe, W. E.,** Dietary pantothenic acid requirement of fingerling channel catfish, *J. Nutr.*, 113, 2124, 1983.
53. **Andrews, J. W. and Murai, T.,** Dietary niacin requirements for channel catfish, *J. Nutr.*, 108, 1508, 1978.
54. **Robinson, E. H. and Lovell, R. T.,** Essentiality of biotin for channel catfish (*Ictalurus punctatus*) fed lipid and lipid-free diets, *J. Nutr.*, 108, 1600, 1978.
55. **Lovell, R. T. and Buston, J. C.,** Biotin supplementation of practical diets for channel catfish, *J. Nutr.*, 114, 1092, 1984.
56. **Kashiwada, K., Kanazawa, A., and Teshima, S.,** Studies on the production of B vitamins by intestinal bacteria. VI. Production of folic acid by intestinal bacteria of carp, *Mem. Fac. Fish, Kagoshima Univ.*, 20, 185, 1971.
57. **Robinson, E. H.,** Channel catfish nutrition, *Rev. Aquatic Sci.*, 1, 365, 1989.
58. **Lovell, R. T.,** personal communication, 1990.
59. **Limsuwan, T. and Lovell, R. T.,** Intestinal synthesis and absorption of vitamin B_{12} in channel catfish, *J. Nutr.*, 111, 133, 1981.
60. **Wilson, R. P. and Poe, W. E.,** Choline nutrition of fingerling channel catfish, *Aquaculture*, 68, 65, 1988.
61. **Burtle, G. J. and Lovell, R. T.,** Lack of response of channel catfish (*Ictalurus punctatus*) to dietary myo-inositol, *Can. J. Fish. Aquat. Sci.*, 46, 218, 1989.
62. **Andrews, J. W. and Murai, T.,** Studies on the vitamin C requirements of channel catfish (*Ictalurus punctatus*), *J. Nutr.*, 105, 557, 1975.
63. **Lim, C. and Lovell, R. T.,** Pathology of the vitamin C deficiency syndrome in channel catfish (*Ictalurus punctatus*), *J. Nutr.*, 108, 1137, 1978.
64. **Lovell, R. T.,** Essentiality of vitamin C in feeds for intensively fed caged channel catfish, *J. Nutr.*, 103, 134, 1973.
65. **Lovell, R. T. and Lim, C.,** Vitamin C in pond diets for channel catfish, *Trans. Am. Fish. Soc.*, 107, 321, 1978.
66. **Wilson, R. P. and Poe, W. E.,** Impaired collagen formation in the scorbutic channel catfish, *J. Nutr.*, 103, 1359, 1973.
67. **Li, Y. and Lovell, R. T.,** Elevated levels of dietary ascorbic acid increase immune responses in channel catfish, *J. Nutr.*, 115, 123, 1985.
68. **Durve, V. S. and Lovell, R. T.,** Vitamin C and disease resistance in channel catfish (*Ictalurus punctatus*), *Can. J. Fish. Aquat. Sci.*, 39, 948, 1982.
69. **Mayer, F. L., Mehrle, P. M., and Crutcher, P. L.,** Interactions of toxaphene and vitamin C in channel catfish, *Trans. Am. Fish. Soc.*, 107, 326, 1978.
70. **Brandt, T. M., Deyoe, C. W., and Seib, P. A.,** Alternate sources of vitamin C for channel catfish, *Prog. Fish-Cult.*, 47, 55, 1985.
71. **Murai, T., Andrews, J. W., and Bauernfeind, J. C.,** Use of L-ascorbic acid, ethocel coated ascorbic acid and ascorbate 2-sulfate in diets for channel catfish, *Ictalurus punctatus*, *J. Nutr.*, 108, 1761, 1978.
72. **Wilson, R. P., Poe, W. E., and Robinson, E. H.,** Evaluation of L-ascorbyl-2-polyphosphate (AsPP) as a dietary ascorbic acid source for channel catfish, *Aquaculture*, 81, 129, 1989.
73. **Robinson, E. H., Rawles, S. D., Brown, P. B., Yette, H. E., and Green, L. W.,** Dietary calcium requirement of channel catfish, *Ictalurus punctatus*, reared in calcium-free water, *Aquaculture*, 53, 263, 1986.
74. **Andrews, J. W., Murai, T., and Campbell, C.,** Effects of dietary calcium and phosphorus on growth, food conversion, bone ash and hematocrit levels of catfish, *J. Nutr.*, 103, 766, 1973.
75. **Lovell, R. T.,** Dietary phosphorus requirement of channel catfish (*Ictalurus punctatus*), *Trans. Am. Fish. Soc.*, 197, 617, 1978.
76. **Wilson, R. P., Robinson, E. H., Gatlin, D. M., III, and Poe, W. E.,** Dietary phosphorus requirement of channel catfish, *J. Nutr.*, 112, 1197, 1982.
77. **Gatlin, D. M., III, Robinson, E. H., Poe, W. E., and Wilson, R. P.,** Magnesium requirement of fingerling channel catfish and signs of magnesium deficiency, *J. Nutr.*, 112, 1181, 1982.
78. **Wilson, R. P.,** unpublished data, 1990.
79. **Gatlin, D. M., III and Wilson, R. P.,** Dietary zinc requirement of fingerling channel catfish, *J. Nutr.*, 113, 630, 1983.
80. **Gatlin, D. M., III and Wilson, R. P.,** Zinc supplementation of practical channel catfish diets, *Aquaculture*, 41, 31, 1984.
81. **Gatlin, D. M., III and Wilson, R. P.,** Dietary selenium requirement of fingerling channel catfish, *J. Nutr.*, 114, 627, 1984.
82. **Gatlin, D. M., III and Wilson, R. P.,** Studies on the manganese requirement of fingerling channel catfish, *Aquaculture*, 41, 85, 1984.
83. **Gatlin, D. M., III and Wilson, R. P.,** Characterization of iron deficiency and the dietary iron requirement for fingerling channel catfish, *Aquaculture*, 52, 191, 1986.

84. **Gatlin, D. M., III and Wilson, R. P.,** Dietary copper requirement of fingerling channel catfish, *Aquaculture*, 54, 277, 1986.
85. **National Research Council,** *Nutrient Requirements of Coldwater Fishes*, National Academy Press, Washington, D.C., 1981.
86. **Hastings, W. H.,** Feed formulation; physical quality of pelleted feed; digestibility, in *Progress in Sport Fisheries Research, 1966*, U.S. Bureau of Sport Fisheries and Wildlife, Washington, D.C., 39, 137, 1966.
87. **Smith, B. W. and Lovell, R. T.,** Digestibility of nutrients in semipurified rations by channel catfish in stainless steel troughs, *Proc. Ann. Conf. S.E. Assoc. Game Fish Comm.*, 25, 452, 1971.
88. **Smith, B. W. and Lovell, R. T.,** Determination of apparent protein digestibility in feeds for channel catfish, *Trans. Am. Fish. Soc.*, 102, 831, 1973.
89. **Cruz, E. M.,** Determination of Nutrient Digestibility in Various Classes of Natural and Purified Feed Materials for Channel Catfish, Ph.D. dissertation, Auburn University, Auburn, AL, 1975.
90. **Brown, P. B., Strange, R. J., and Robbins, K. R.,** Protein digestibility coefficients for yearling channel catfish fed high protein feedstuffs, *Prog. Fish-Cult.*, 47, 94, 1985.
91. **Wilson, R. P. and Poe, W. E.,** Apparent digestible protein and energy coefficients of common feed ingredients for channel catfish, *Prog. Fish-Cult.*, 47, 153, 1985.
92. **Wilson, R. P., Robinson, E. H., and Poe, W. E.,** Apparent and true availability of amino acids from common feed ingredients for channel catfish, *J. Nutr.*, 111, 923, 1981.
93. **Dupree, H. K.,** Feeding practices, in *Nutrition and Feeding of Channel Catfish*, Revised, Robinson, E. H., and Lovell, R. T., Eds., Cooperative Series Bull. 296, Texas A&M Univ., College Station, 1984.

COMMON CARP, *CYPRINUS CARPIO*

Shuichi Satoh

INTRODUCTION

Common carp originated in eastern Europe and central Asia. Farming of common carp started in European countries and Japan many centuries ago, and has been eventually introduced to other parts of the world. More than 85% of the carp production takes place in eastern Europe, and estimates indicate that production has doubled during the last decade to the current level of approximately 430,000 t in 1985, of which the USSR contributes more than 60%. Other major carp producing countries are Japan and Indonesia.[1]

Although carp has been farmed since ancient times, scientific study of their nutrition is of recent origin. The first serious studies were made immediately postwar and the subject has gathered pace since the mid-1960s. Most of the research on the nutritional requirements of common carp has been conducted on small fish under laboratory conditions and on postjuvenile stages in net-cage culture with practical diets.

NUTRIENT REQUIREMENTS

The growth rate, feed conversion, and carcass composition of fish may be affected by species, genetic strain, sex, stage of reproductive cycle, etc., leading to different nutritional requirements. Growth rates may also be affected by diet quality in terms of nutrient balance, energy content, and nutrient bioavailability, as well as environmental conditions, such as water temperature, oxygen content, water flow rate, etc. The total requirement for a given nutrient during growth includes the amount needed for maintenance, as well as the amount required for the formation of new tissues. The values given in the feeding tables of this review represent these combined requirements. Of the various nutrient needs for growth, the requirement for energy is by far the largest and primarily governs the total food allowance.

Thus, all types of formulated fish feed must satisfy the nutritional requirement of the cultured species in terms of protein (essential amino acids [EAA]), lipid (essential fatty acids [EFA]), energy, vitamins, and minerals. The quality of the feed ultimately will depend upon the level of nutrients available to the fish. However, it is difficult to define the nutritional quality of feeds because of the interactions that occur between various nutrients during and after digestion and absorption. This is also greatly affected if the fish eat to satisfy their need for energy. Thus, the actual nutrient intake is regulated by the available energy level of the diet and the energy requirement of the fish.

PROTEIN AND AMINO ACIDS

When proteins are hydrolyzed, about 20 different amino acids are obtained. Consequently, the study of protein nutrition deals primarily with the quantity and bioavailability of these amino acids. The quality of protein depends on its digestibility, biological value, and net protein utilization.

Studies on the essential amino acid requirements of common carp have demonstrated that they require the same 10 essential amino acids as other fishes. These are arginine, histidine, isoleucine, leucine, lysine, methionine, phenylalanine, threonine, tryptophan, and valine (Table 1).[2,3]

The quality of protein is usually evaluated by both biological and chemical methods; however, the former, in which body weight gain and nitrogen retention are used as criteria for

TABLE 1
Essential Amino Acid Requirements
of Common Carp[a]

Amino acid	Ogino[2]	Nose[3]
Arginine	4.4	4.2
Histidine	1.5	2.1
Isoleucine	2.6	2.3
Leucine	4.8	3.4
Lysine	6.0	5.7
Methionine	1.8	3.1[b]
Cystine	0.9	—
Phenylalanine	3.4	6.5[c]
Tyrosine	2.3	—
Threonine	3.8	3.9
Tryptophan	0.8	0.8
Valine	3.4	3.6

[a] Expressed as percent of dietary protein.
[b] Expressed as methionine plus cystine.
[c] Expressed as phenylalanine plus tyrosine.

protein quality, is more accurate. In the biological methods, protein efficiency ratio, biological value, and net protein utilization (NPU) are most frequently utilized. Ogino et al.[4] reported that NPU can be calculated using the following formula:

$$\text{NPU} = \frac{\text{True nitrogen retained}}{\text{Nitrogen intake}} \times 100$$

$$= \frac{\begin{array}{c}\text{Nitrogen increase in fish} \\ \text{fed the test protein diet}\end{array} + \begin{array}{c}\text{Nitrogen decrease in fish} \\ \text{fed the protein free diet}\end{array}}{\text{Nitrogen intake from the test protein diet}} \times 100$$

Nitrogen decrease in common carp fed the protein free diet
= (metabolic fecal nitrogen + endogenous nitrogen) × days
$$= \frac{\text{Initial + final}}{2} \text{ (total body wt. g)} \times 0.00014 \text{ g} \times \text{days}$$

where

metabolic fecal nitrogen[5] = 0.00003 g × body wt. g × days
endogenous nitrogen[5] = 0.00011 g × body wt. g × days.

The protein requirements of fish are influenced by various factors, such as fish size, water temperature, feeding rate, availability and quality of natural foods, overall digestible energy content of the diet, and the quality of the protein. The protein requirement of fish is closely related to the optimum dietary protein level, although the two are not identical. The former value is characteristic for each species under controlled rearing conditions and should be expressed as a value per unit of fish body weight per day. Ogino and Chen[6] and Ogino[7] reported that the requirement of common carp for protein is about 1 g/kg body weight/d for maintenance and 12 g/kg body weight/d for maximum protein retention. The efficiency of nitrogen utilization for growth, however, is highest with a protein intake of 7 to 8 g/kg body weight/d. Consequently, the optimal protein level in practical feeds is based on the protein requirements, bearing in mind the quality of the protein, the digestible energy levels, the rearing conditions, the manufacturing process for the feed, and market costs.

Investigations on the optimal protein requirement of common carp have demonstrated that levels of 30 to 38% crude protein in the diet appear to satisfy the fish.[8] Generally, this protein level has been determined by using semipurified diets containing single high-quality protein sources, such as casein, whole-egg protein, or fish meal. If sufficient digestible energy is contained in the diet, the optimal protein level can be held at 30 to 35%.[9]

The protein requirement of common carp was found to be very similar to that of rainbow trout and swine, but lower than that of broiler chicken when expressed as g/kg body weight/d.[7] This is due to the fact that since fish are poikilotherms, their feeding rate is significantly lower than that of homoisotherms.

LIPIDS AND ESSENTIAL FATTY ACIDS

Fish utilize lipids for energy, cellular structure, and maintenance of the integrity of biomembranes. The fluidity of biomembranes is regulated partly by the fatty acids contained in the phospholipids that control such processes as cellular transport and the activity of membrane-associated enzymes. Most fish are different from terrestrial animals in that their tissues contain fairly high amounts of n-3 highly unsaturated fatty acids (n-3 HUFA) such as 20:5n-3 and 22:6n-3. The degree of unsaturation in tissues increases when environmental temperature is lowered, thereby maintaining membrane fluidity to allow normal cellular function.

In addition to serving as an important source of energy, dietary lipids serve as the source of essential fatty acids and as a carrier for the fat-soluble vitamins A, D, E, and K. Studies on lipid nutrition of fish have shown that the requirements for EFA differ considerably from species to species, most notably between freshwater and marine fish.[9]

Initial experiments using carp fingerlings (2.5 g) fed a fat-free diet for a fairly long time were unsuccessful in inducing an EFA deficiency, quite unlike rainbow trout.[10] Adding saturated lipid (5% methyl laurate) resulted in a positive growth response, but supplements of 18:2n-6 or 18:3n-3 in the diet resulted in very little improvement in growth and feed efficiency after 22 weeks. The relatively low growth rate observed in fish fed the fat-free diet may be attributed to their lower caloric intake. Watanabe et al.[11] and Takeuchi and Watanabe[12] conducted feeding trials using very young carp weighing about 0.65 g that had been kept on a fat-free diet for 4 months before the initiation of the feeding trial, and concluded that these fish had an EFA requirement for both 18:2n-6 and 18:3n-3. The best weight gain and feed efficiency were obtained in fish receiving a diet containing 1% of 18:2n-6 and 1% of 18:3n-3.

Body lipid deposition in common carp is directly related to dietary lipid levels.[13-15] Although feeding high-quality lipids ranging from 5 to 25% of the diet never led to ill effects in common carp, there was a marked increase in visceral fat due to the excess energy density of the diets. However, the lipid content of hepatopancreas was not affected by dietary lipid levels.

Dietary lipids lacking EFA or containing oxidation products can, however, affect body composition. Castell et al.[16] reported increased moisture content of muscle in rainbow trout fed the EFA-deficient diet, apart from higher liver respiration rates and lower blood hemoglobin. Watanabe et al.[11] confirmed that the whole body, muscle, and viscera of fish fed EFA-deficient diets were considerably higher in moisture and lower in both protein and lipid content. Feeding the EFA-deficient diet resulted in a decrease in the lipid content of the hepatopancreas in common carp.[12] These changes in lipid content were found to be due to alterations in neutral lipids; the content of tissue polar lipids remained at an almost constant level, apparently unaffected by the EFA deficiency, and the level of nonpolar lipids in the liver increased in nonessential fatty acids, such as 12:0 and 16:0.[17] The accumulation of lipid in livers of EFA-deficient animals has been suggested to be due to an impairment of lipoprotein biosynthesis.[18,19]

The fatty acid composition of body lipids reflects the composition of dietary lipids. Fish are able to alter the dietary n-6/n-3 ratio in favor of n-3 fatty acid incorporation into body lipids,[20] but the influence of dietary lipids on the fatty acid composition of body lipids differs between triglycerides and phospholipids. Dietary lipids affect the fatty acid composition of phospholipids to a greater degree than those in the neutral lipids. In freshwater fish, dietary 18:2n-6 and 18:3n-3 are elongated and desaturated, and the former is converted to 20:4n-6 and 22:5n-6, while the latter to 22:6n-3 in the phospholipid fraction; whereas, in the triglyceride fraction, these fatty acids are deposited unaltered, increasing the concentration of 18:2n-6 and 18:3n-3.[12,21] Farkas et al.[22] suggested that there is a constant flow of linolenate from triglycerides, via chain elongation and desaturation, to the different phospholipids in carp. As long as the diet contains sufficient linolenate, this fatty acid will either be deposited in triglycerides unchanged or converted to 22:6n-3, and incorporated in phospholipids and triglycerides.

The EFA status of diets is known to affect the fatty acid composition of phospholipids in fish. Feeding a fat-free diet or an EFA-deficient diet results in elevated levels of eicosatrienoic acid (20:3n-9), along with an elevation of monoethlenic fatty acids in phospholipids and 18:1n-9 in triglycerides in carp. Dietary 18:2n-6 and 18:3n-3 both depress the triene levels. The appearance of 20:3n-9 in phospholipids is somehow related to the disappearance of 18:3n-3 and 22:6n-3 from the triglycerides, and the accumulation of 20:3n-9 appears after the linolenate pool is depleted in the triglycerides.

The increase of 20:3n-9 in the tissue of fish fed EFA-deficient diets agrees with the findings of other workers for the rat,[23] chick,[24] and rabbit.[25] For most animal studies in which fatty acids of the linoleic type satisfy the EFA requirement, Halman and coworkers[23] have suggested that the ratio of 20:3n-9/20:4n-6 in the tissue lipid can be used as an index of EFA deficiency. Castell et al.[26] suggested, however, that since rainbow trout require fatty acids of the n-3 type, the ratio of 20:3n-9/22:6n-3 would be more appropriate as an index of EFA status for fish. Alfin-Slater and Aftergood[27] proposed the ratio of 20:3n-9/20:5n-3 as an estimate of the adequacy of linolenate intake in the diet, but in the case of trout, 22:6n-3 appears to be the primary long-chain fatty acid of this series that accumulates in body lipids.[26] If the ratio of 20:3n-9/22:6n-3 in the phospholipids is used as a criterion for evaluating EFA deficiency, all fish receiving 0.7% or more of 18:3n-3 in the diet would receive a sufficient amount of n-3 fatty acids. This judgment is based on a ratio of 0.4 or greater indicating a deficiency in mammals. In common carp, which require fatty acids of both 18:2n-6 and 18:3n-3 for growth, Watanabe et al.[11] proposed the ratio of 20:3n-9/20:4n-6, as well as the ratio of 20:3n-9/22:6n-3 for the EFA index in carp. When both ratios were applied to carp as indices, the former ratio became less than 0.6 and the latter less than 0.4 if the carp were receiving a sufficient amount of both n-6 and n-3 fatty acids.

The main protein-sparing effect of dietary lipids is to replace protein that would have otherwise been catabolized and used both for energy and to synthesize lipid.[28] The sparing of dietary protein by lipids has been extensively investigated in common carp.[13-15]

Omnivorous fish, such as common carp, can utilize both carbohydrate and lipid effectively as dietary energy sources. An increase of the digestible energy (DE) content from 320 to 460 kcal/100 g diet by the addition of 5 to 15% lipid to 32% crude protein diets did not improve growth or feed conversion. In fish fed diets containing 23% protein, values for these parameters were quite low, regardless of the DE content in the diet.[15] The optimum DE/protein ratio for growth of carp was found to be 97 to 116 when based on measured DE values.

Animal fats, such as lard and tallow, or hydrogenated beef tallow, along with other lipids providing the necessary level of EFA, can be used as an energy source for certain fish diets without any adverse effects on feed efficiency, fish growth, or survival.[29] Takeuchi et al.[13,30] determined the digestibility of beef tallow and various hydrogenated fish oils with different melting points in common carp and also investigated the effects of fish size and water temperature on the digestibility of these lipids. The digestibility of hydrogenated fish oils was found to be affected by their melting point (mp) and increased as melting point decreased.

Whereas the digestibility of dietary protein was as high as 98% without regard to that of the lipid. The hydrogenated oils of mp 53°C was significantly lower in digestibility in carp, especially in fish weighing less than 10 g. On the other hand, beef tallow and hydrogenated fish oil of mp 38°C were found to be effectively utilized by carp, with a digestibility of more than 70%, regardless of fish size and water temperature.

High intake of dietary lipid containing polyenoic fatty acids may increase the requirement for vitamin E. A relationship between the α-tocopherol requirement and the polyunsaturated fatty acid (PUFA) content of the diet has been reported by many workers. Watanabe et al.[31,32] investigated whether common carp had an increased requirement for dietary tocopherol as the unsaturation of the fat in the diet was increased. In their experiment, they observed the so-called Sekoke disease characterized by a marked loss of flesh (muscular dystrophy) in the back, which is known to be induced by the absence of α-tocopherol[33-35] or oxidized lipids.[36-38] They demonstrated that the elevated dietary linoleate levels increased the α-tocopherol requirement of carp. This was confirmed in carp by feeding diets containing 15 to 20% pollock liver oil, which resulted in an increased requirement for α-tocopherol.[39]

Oxidized lipids are also known to be toxic for various fish. Hashimoto et al.[36] demonstrated that Sekoke disease, a serious problem for common carp culturists in Japan, was induced by oxidized lipids in dried silkworm pupae that was used as a main feedstuff. They found that the disease was effectively prevented by the addition of DL-α-tocopheryl acetate (50 mg/100 g diet), but not by antioxidants, such as methylene blue, BHA, ethyl gallate, DPPD, and ethoxyquin.[37,40] The result of the preventive effect of α-tocopherol and ethoxyquin obtained by Watanabe et al.[36-40] and Murai and Andrews[41] is in concurrence with that of Takeuchi,[42] who reported that α-tocopherol may act as a peroxide decomposer and stabilizing factor of intestinal epithelium, which improves the absorption of oxidized lipids; while ethoxyquin did not improve absorption of hydroperoxide. The preventive effect of tocopherol homologues on muscular dystrophy in young common carp has been investigated by Aoe et al.,[43] who reported that α-tocopherol was far more effective than β-, γ-, and δ-tocopherols, in the order of decreasing potency. Probably in common carp, the assimilation of non-α-tocopherols is very low.[44]

The low content of tissue tocopherol in carp fed diets containing oxidized lipids suggests that the muscular dystrophy induced by oxidized lipid was coupled with an α-tocopherol deficiency.[37,40] It is most likely that the oxidized lipids destroyed the α-tocopherol originally present in the diet. Watanabe et al.[32-35,39] proved that the α-tocopherol-deficient diet resulted in an apparent muscular dystrophy in common carp after 90 d of feeding and that the requirement of common carp for α-tocopherol to support optimal growth was about 10 mg/100 g diet, at a dietary lipid level of 5%. Care should be taken not to use oxidized lipids, and diets should contain antioxidants to prevent further oxidation.

CARBOHYDRATES

Carbohydrates can act either as an immediate energy source or as rapidly available energy reserves stored as glycogen in the liver and muscle. Many studies on carbohydrate utilization have been conducted with common carp and other fishes. It is generally accepted that the ability of fish to utilize carbohydrates is lower than terrestrial animals, and the utilization of carbohydrate differs among fish species. The intestinal activity of amylase, which hydrolyzes starch, is higher in omnivorous fishes, such as common carp, than in carnivorous fishes. It has been shown that the digestibility of gelatinized starch in common carp is high and unaffected by dietary level. In addition, the amylase activity in the digestive tract and the digestibility of starch in fishes are generally lower than that of humans and terrestrial animals. On the other hand, the digestibility of dextrin is comparatively high in certain fish.[45] Also, glucose is rapidly absorbed from the gut tract of fish.[46]

In general, fish exhibit poorer growth and feed efficiency values when fed high levels of dietary carbohydrate.[45,47] Common carp utilize gelatinized starch more effectively than dextrin and glucose. Furuichi and Yone[46] reported that the growth and feed efficiency of carp fed diets containing gelatinized starch at a level of 42% was higher than for fish fed diets containing either dextrin or glucose at the same level. Murai et al.[48] investigated the effects of various dietary carbohydrates and the frequency of feeding on their respective utilization by fingerling carp. At twice daily feeding, the starch diet resulted in the best weight gain and feed efficiency. However, carp were found to utilize glucose and maltose as effectively as starch when fed at least four times daily. The improvement of glucose or maltose utilization by carp with the increase in feeding frequency was considered to be due to a reduction in the rapid absorption of large amounts of the glucose being fed.

The optimum level of dietary carbohydrate for fishes is generally low and varies among species. Ogino et al.[49] found that common carp utilized carbohydrate effectively as an energy source. Takeuchi et al.[14] reported that both carbohydrate and lipid have good nutritive value for common carp as dietary energy sources. However, it is very difficult to estimate the optimum level of dietary carbohydrate for fish, because protein and lipid are also used as energy sources, and the utilization of carbohydrate may be affected by the levels of dietary protein and lipid. Based on the results of many studies, the optimum level of dietary carbohydrate may be considered to be 30 to 40% for common carp.

VITAMINS

Experiments on the qualitative and quantitative vitamin requirements of common carp have shown that two fat-soluble and eight water-soluble vitamins are required. No deficiency signs were observed when fish were fed diets devoid in vitamins B_{12}, C, D, K, and para-aminobenzoic acid. Vitamin requirement values and their respective deficiency signs are summarized in Table 2.

Vitamin requirements of common carp may be affected by various factors, such as fish size, water temperature, and diet composition. For example, the requirement for vitamin E may increase as the level of PUFA in the diet increases. In extensive culture and semiintensive culture in ponds or lakes, natural live food is often sufficiently abundant to provide for the essential vitamins. However, in intensive high-density culture, such as in heavily stocked ponds, or in cages and raceways, natural food is scarce, demanding a supply of vitamins in the diet to achieve normal growth.

Certain vitamins may be destroyed during feed manufacture, by heat, moisture, alteration of pH, the presence of some metals, lipid oxidation, etc. Supplemental levels of each vitamin in fish feeds are recommended to be higher than the requirement values in order to provide a safety margin.

MINERALS

Minerals required by common carp are magnesium, phosphorus, and a number of trace elements, such as copper, cobalt, iron, manganese, and zinc. Mineral requirement values and their deficiency signs are summarized in Table 3. Of all the minerals, phosphorus is one of the most important, mainly because it is required for growth and bone mineralization, and also for lipid and carbohydrate metabolism. The level of dietary phosphorus required is much higher than the other inorganic ions, with dietary levels of 0.6 to 0.7% available phosphorus being required for common carp.[60,61] Generally, phosphorus requirements are not affected by dietary calcium levels in fish. In controlled experiments, the growth of carp has been shown to correlate positively with dietary phosphorus levels but not with calcium levels. However, for most fish species it is difficult to study the effects of calcium deficiency because calcium is readily absorbed from the water via the gills.

The availability of inorganic phosphorus depends on the solubility of the inorganic salt. Thus, phosphorus from tricalcium phosphate is less available than that from the more soluble

TABLE 2
Vitamin Requirements and Deficiency Signs of Common Carp

Vitamin	Requirement (mg/kg)	Deficiency signs
Thiamine	R[50]	Poor growth
Riboflavin	7–14[51,52]	Anorexia, poor growth, hemorrhages in hepatopancreas and skin
Pyridoxine	5–6[53]	Anorexia, ascites, ataxia, exophthalmia, convulsions
Pantothenate	30–50[51]	Anorexia, poor growth, irritability, hemorrhages in skin
Niacin	28[54]	Anorexia, poor growth, poor survival, hemorrhages in skin
Biotin	1[55]	Poor growth, erythrocyte fragility and fragmentation
Choline	4000[56]	Poor growth, fatty hepatopancreas
Inositol	440[57]	Anorexia, poor growth, dermatitis
Vitamin A	10,000 IU[58]	Anorexia, poor growth, exophthalmia, hemorrhage in gill and skin, discoloration of skin, opercular malformation
Vitamin E	100[32,39]	Muscular dystrophy
Vitamin C	R[59]	Caudal fin erosion and deformed gill arches in larval stage

mono- and dicalcium phosphates, particularly in stomachless fish such as common carp.[61] It has also been shown that the availability of phosphorus contained in fish meal is fairly low in common carp (0 to 33%) compared with rainbow trout (60 to 81%).[61,66] Supplementation of monosodium phosphate to fish meal diets resulted in an increase in growth response of common carp.[67,68] This difference in phosphorus availability from fish meal between carp and rainbow trout is considered to be due to the presence of the acidic gastric juices in the stomach of the trout. Phosphorus in fish meal exists mainly in the form of insoluble hydroxyapatite $[Ca_{10}(PO_4)_6(OH)_2]$ originating from the hard tissue, such as bones and scales. Accordingly, it is presumed that common carp cannot dissolve the phosphorus in fish meal, and consequently require supplemental soluble phosphorus in the diet.

Satoh et al.[69] determined the effect of deletion of magnesium from the mineral supplement in a white fish meal diet on growth and mineral composition of vertebrae in common carp. The basal diet contained 0.14% magnesium derived from white fish meal, or about twice the level required by common carp. Feeding the basal diet resulted in poor growth and greatly affected the mineral composition of the vertebrae. The ratio of calcium to magnesium in the vertebrae of fish fed the white fish meal diet without supplemental magnesium was greater than 50. A rate of 50 or less has been reported to indicate normal status of fish receiving adequate levels of dietary magnesium. Thus, the magnesium in fish meal is poorly absorbed by common carp.

Satoh et al.[69] conducted a long-term feeding study to determine the availability of various trace minerals contained in fish meal to carp. Feeding a fish meal diet without supplemental manganese resulted in lens cataracts (70% of fish), along with short-body dwarfism (90%) and depressed growth. Deletion of zinc, copper, or cobalt from the mineral supplement also resulted in cataracts (40 to 70%) and reduced growth.

Satoh et al.[70] also examined the availability of manganese contained in white fish meal and the minimum supplemental manganese levels needed for a white fish meal-based diet for normal growth of common carp. The lowest growth rate and the highest incidence of dwarfism

TABLE 3
Mineral Requirements and Deficiency Signs of
Common Carp

Mineral	Requirement	Deficiency signs
Phosphorus	0.6–0.7%[60,61]	Poor growth, skeletal abnormalities, low feed efficiency, low ash in whole body and vertebrae, high lipid content
Magnesium	0.04–0.05%[62]	Poor growth, high mortality, sluggishness and convulsions, high calcium content in bone
Zinc	15–30 mg/kg[63]	Poor growth, high mortality, erosion of fins and skin, low zinc and manganese content in bone
Manganese	13 mg/kg[65]	Poor growth, dwarfism, skeletal abnormalities, high mortality, low calcium, magnesium, phosphorus, zinc and manganese in bone
Copper	3 mg/kg[65]	Poor growth
Cobalt	0.1 mg/kg	Poor growth
Iron	150 mg/kg[64]	Low specific gravity, hemoglobin content, and hematocrit values, abnormal mean corpuscular diameter

were observed in fish receiving a diet without supplemental manganese. Performance was effectively improved by the addition of $MnCl_2$ at a level greater than 10 mg Mn/kg diet. The diet without supplemental manganese contained 3 mg Mn/kg diet, derived mostly from white fish meal. The addition of 10 mg Mn/kg diet resulted in a total Mn level of 13 mg/kg diet, equivalent to the manganese requirement of carp as determined by Ogino and Yang[65] using semipurified diets. This level resulted in satisfactory performance in terms of growth, feed efficiency, and absence of dwarfism. These results indicate a high availability of manganese in white fish meal to carp. Based on Mn availability, among the various manganese compounds ($MnSO_4$, $MnCO_3$, MnO_2, $MnCl_2$), $MnSO_4$ and $MnCl_2$ were the most suitable sources of manganese for common carp. The availability of Mn from MnO_2 and $MnCo_3$ was very low.

Satoh et al.[71] reported that the manganese contained in four kinds of fish meal (white fish meal, brown fish meal, and sardine meal with or without soluble) was found to be highly available to carp.

PRACTICAL DIETS

Watanabe et al.[72] conducted field experiments to develop practical carp diets that were effective in reducing nitrogen excretion from common carp without reduction of growth or feed efficiency, and an increase of feed cost. The experimental diets were prepared by using locally available low-cost ingredients and were fed to common carp in net cages in Lake Kasumigaura in Japan (Table 4). They reduced the dietary protein level from 39%, the official standard for common carp feeds in Japan, to 30 to 35% by increasing the energy level with carbohydrate, such as wheat flour, or with lipid, such as beef tallow. The use of practical ingredients in the experiment resulted in a reduction of feed costs. The most suitable dietary protein and digestible energy levels for carp cultured in Lake Kasumigaura to reduce nitrogen excretion were found to be about 35% and 350 kcal/100 g diet, respectively. Feeding this low

TABLE 4
**Composition of Practical Type Diets for Common Carp Along
with Growth after 20 Weeks of Feeding and Total Nitrogen
Excreted by the Fish[72]**

Ingredient	1	2	3	4	5	6
Brown fish meal	25	25	15	30		
Meat meal	4	4	10	4		
Soybean meal	6	10	12	10		
Defatted wheat					Commercial	
germ meal	0	5	0	5		
Corn gluten meal	5	5	10	8		
Wheat middlings	42	37	33	29		
Defatted rice						
bran	9.5	9.5	9.5	9.5	diet	
NaH_2PO_4	2.5	2.5	2.5	2.5		
Mineral mixture	0.5	0.5	0.5	0.5		
Vitamin mixture	0.5	0.5	0.5	0.5		
Beef tallow	5	0	4	0		
Soybean oil	5	1	1	1		
Crude protein	30.7	34.7	36.2	39.4	35.6	40.2
Crude lipid	12.5	8.4	11.7	8.9	7.5	8.0
Gross energy						
(kcal/100 g)	450	420	440	430	420	420
Growth rate						
(20 weeks, %)	775	829	804	751	688	747
Feed efficiency	0.80	0.76	0.80	0.83	0.80	0.73
Nitrogen excretion						
kg/ton production	37.4	48.1	47.1	51.4	45.8	64.7

protein-high energy diet resulted in a reduction in the total nitrogen excretion per ton of common carp produced by 22 to 36% of the value observed when commercial diets were fed.

Since common carp is a fish without a stomach, it is better to feed them frequently. Kariya[73] conducted a study to determine the effect of feeding frequency on growth of juvenile common carp at 27 to 32°C. Growth rates reached a plateau when the diet was fed more than five times a day. Tominaga[74] developed the following formula to calculate the feeding rate per day for common carp:

Feeding rate per day = water temperature(°C) × 0.13% (dry basis).

Kurihara[75] reported that feeding rate was affected strongly by water temperature and fish size in net-cage culture of carp (Table 5). He recommended a table of feeding rate that could be used for 50 to 900 g carp by feeding four times a day. Tazaki[76] also reported a similar table (Table 6) for juvenile carp (less than 50 g).

TABLE 5
Feeding Rate for Cage Culture of Common Carp per Day[75]

Water temperature (°C)	Body weight in g					
	50–100	100–200	200–300	300–700	700–800	800–900
15	2.4	1.9	1.6	1.3	1.1	0.8
16	2.6	2.0	1.7	1.4	1.1	0.8
17	2.8	2.2	1.8	1.5	1.2	0.9
18	3.0	2.3	1.9	1.7	1.3	1.0
19	3.2	2.5	2.0	1.8	1.4	1.0
20	3.4	2.7	2.2	1.9	1.5	1.1
21	3.6	2.9	2.3	2.0	1.6	1.2
22	3.9	3.1	2.5	2.2	1.7	1.3
23	4.2	3.3	2.7	2.3	1.8	1.4
24	4.5	3.5	2.9	2.5	2.0	1.5
25	4.8	3.8	3.1	2.7	2.1	1.6
26	5.2	4.1	3.3	2.9	2.3	1.7
27	5.5	4.4	3.5	3.1	2.4	1.8
28	5.9	4.7	3.8	3.3	2.6	1.9
29	6.3	5.0	4.1	3.5	2.8	2.1
30	6.8	5.4	4.4	3.8	3.0	2.2

TABLE 6
Feeding Rate for 1-Year-Old Common Carp Per Day[76]

Water temperature (°C)	Body weight in g					
	2–5	5–10	10–20	20–30	30–40	40–50
15	4.9	4.1	3.3	3.1	2.7	2.2
16	5.2	4.4	3.5	3.3	2.9	2.3
17	5.5	4.7	3.7	3.6	3.1	2.5
18	5.8	5.0	4.0	3.9	3.4	2.7
19	6.3	5.4	4.4	4.2	3.7	2.9
20	6.9	5.9	4.9	4.6	4.0	3.2
21	7.5	6.4	5.2	4.9	4.3	3.4
22	8.1	6.9	5.6	5.3	4.5	3.6
23	8.7	7.4	6.0	5.6	4.9	3.9
24	9.2	7.9	6.4	6.0	5.1	4.1
25	9.8	8.2	6.7	6.2	5.4	4.4
26	10.4	8.8	7.0	6.6	5.8	4.6
27	11.0	9.4	7.5	7.2	6.2	5.0
28	11.6	10.0	8.1	7.8	6.8	5.4
29	12.6	10.8	8.9	8.4	7.4	5.8
30	13.8	11.8	9.8	9.2	8.0	6.4

REFERENCES

1. **Ackefors, H.,** Europe: the emerging force in aquaculture, *World Aquaculture*, 19, 5, 1988.
2. **Ogino, C.,** Requirements of carp and rainbow trout for essential amino acids, *Bull. Jpn. Soc. Sci. Fish.*, 46, 171, 1980.
3. **Nose, T.,** Summary report on the requirements of essential amino acids for carp, in *Finfish Nutrition and Fishfeed Technology*, Tiews, K. and Halver, J. E., Eds., Heenemann, Berlin, 1979, 145.
4. **Ogino, C., Kawasaki, H., and Nanri, H.,** Method for the determination of nitrogen retained in the fish body by the carcass analysis, *Bull. Jpn. Soc. Sci. Fish.*, 46, 105, 1980.

5. **Ogino, C., Kakino, J., and Chen, M.-S.,** Protein nutrition in fish-II. Determination of metabolic fecal nitrogen and endogenous nitrogen excretions of carp, *Bull. Jpn. Soc. Sci. Fish.*, 39, 519, 1973.
6. **Ogino, C. and Chen, M.-S.,** Protein nutrition in fish-V. Relation between biological value of dietary proteins and their utilization in carp, *Bull. Jpn. Soc. Sci. Fish.*, 39, 955, 1973.
7. **Ogino, C.,** Protein requirement of carp and rainbow trout, *Bull. Jpn. Soc. Sci. Fish.*, 46, 385, 1980.
8. **Watanabe, T.,** Nutrition and growth, in *Intensive Fish Farming*, Shepherd, C. J. and Bromage, N. R., Eds., BSP Professional Books, London, 1988, 154.
9. **Watanabe, T.,** Lipid nutrition in fish, *Comp. Biochem. Physiol.*, 73A, 3, 1982.
10. **Watanabe, T., Utsue, O., Koshiishi, Y., and Ogino, C.,** Effect of dietary methyl linoleate and linolenate on growth of carp-I, *Bull. Jpn. Soc. Sci. Fish.*, 41, 257, 1975.
11. **Watanabe, T., Takeuchi, T., and Ogino, C.,** Effect of dietary linoleate and linolenate on growth of carp-II, *Bull. Jpn. Soc. Sci. Fish.*, 41, 263, 1975.
12. **Takeuchi, T. and Watanabe, T.,** Requirement of carp for essential fatty acids, *Bull. Jpn. Soc. Sci. Fish.*, 43, 541, 1977.
13. **Takeuchi, T., Watanabe, T., and Ogino, C.,** Use of hydrogenated fish oil and beef tallow as dietary energy source for carp and rainbow trout, *Bull. Jpn. Soc. Sci. Fish.*, 44, 875, 1978.
14. **Takeuchi, T., Watanabe, T., and Ogino, C.,** Availability of carbohydrate and lipid as dietary energy sources for carp, *Bull. Jpn. Soc. Sci. Fish.*, 45, 977, 1979.
15. **Takeuchi, T., Watanabe, T., and Ogino, C.,** Optimum ratio of dietary energy to protein for carp, *Bull. Jpn. Soc. Sci. Fish.*, 45, 983, 1979.
16. **Castell, J. D., Sinnhuber, R. O., Wales, J. H., and Lee, J. D.,** Essential fatty acids in the diet of rainbow trout (*Salmo gairdneri*): Physiological symptoms of EFA deficiency, *J. Nutr.*, 102, 87, 1972.
17. **Watanabe, T. and Takeuchi, T.,** Evaluation of pollock liver oil as a supplement to diets for rainbow trout, *Bull. Jpn. Soc. Sci. Fish.*, 42, 893, 1976.
18. **Fukuzawa, T., Privett, O. S., and Takahashi, Y.,** Effect of essential fatty acid deficiency on release of triglycerides by the perfused rat liver, *J. Lipid Res.*, 11, 522, 1970.
19. **Fukuzawa, T., Privett, O. S., and Takahashi, Y.,** Effect of essential fatty acid deficiency on transport from liver, *Lipids*, 6, 388, 1971.
20. **Sinnhuber, R. O.,** The role of fats, in *Fish in Research*, Neuhaus, O. W. and Halver, J. E., Eds., Academic Press, New York, 1969, 245.
21. **Farkas, T., Csenger, I., Majoros, F., and Olah, J.,** Metabolism of fatty acids in fish. III. Biosynthesis of fatty acids in relation to diet in the carp, *Cyprinus carpio* Linnaeus 1758, *Aquaculture*, 20, 29, 1980.
22. **Farkas, T., Csenger, I., Majoros, F., and Olah, J.,** Metabolism of fatty acids in fish. I. Development of essential fatty acid deficiency in the carp, *Cyprinus carpio* Linnaeus 1758, *Aquaculture*, 11, 147, 1977.
23. **Mohrhauer, H. and Holman, R. T.,** Effect of linolenic acid upon the metabolism of linoleic acid, *J. Nutr.*, 81, 67, 1963.
24. **Hill, E. G.,** Effect of dietary linoleate on chick liver fatty acids: dietary linoleate requirement, *J. Nutr.*, 89, 465, 1966.
25. **Ahluwalia, B. G., Pincus, G., and Holman, R. T.,** Essential fatty acid deficiency and its effect upon reproductive organs of male rabbits, *J. Nutr.*, 92, 205, 1967.
26. **Castell, J. D., Lee, J. D., and Sinnhuber, R. O.,** Essential fatty acids in the diet of rainbow trout (*Salmo gairdneri*): lipid metabolism and fatty acid composition, *J. Nutr.*, 102, 93, 1972.
27. **Alfin-Slater, R. B. and Aftergood, L.,** Essential fatty acids reinvestigated, *Physiol. Rev.*, 48, 758, 1968.
28. **Cowey, C. B. and Sargent, J. R.,** Fish nutrition, *Adv. Mar. Biol.*, 10, 383, 1972.
29. **Yu, T. C., Sinnhuber, R. O., and Putnam, G. B.,** Effect of dietary lipids on fatty acid composition of body lipid in rainbow trout (*Salmo gairdneri*), *Lipids*, 12, 495, 1977.
30. **Takeuchi, T., Watanabe, T., and Ogino, C.,** Digestibility of hydrogenated fish oil in carp and rainbow trout, *Bull. Jpn. Soc. Sci. Fish.*, 45, 1521, 1979.
31. **Watanabe, T. and Takashima, F.,** Effect of α-tocopherol deficiency on carp-VI. Deficiency symptoms and changes of fatty acid and triglyceride distributions in adult carp, *Bull. Jpn. Soc. Sci. Fish.*, 43, 819, 1977.
32. **Watanabe, T., Takeuchi, T., Ogino, C., and Kawabata, T.,** Effect of α-tocopherol deficiency on carp-VII. The relationship between dietary levels of linoleate and α-tocopherol requirement, *Bull. Jpn. Soc. Sci. Fish.*, 43, 935, 1977.
33. **Watanabe, T., Takashima, F., Ogino, C., and Hibiya, T.,** Effects of α-tocopherol deficiency on carp, *Bull. Jpn. Soc. Sci. Fish.*, 36, 623, 1970.
34. **Watanabe, T., Takashima, F., Ogino, C., and Hibiya, T.,** Requirement of young carp for α-tocopherol, *Bull. Jpn. Soc. Sci. Fish.*, 36, 972, 1970.
35. **Watanabe, T., Takashima, F., Ogino, C., and Hibiya, T.,** Effects of α-tocopherol deficiency on carp-II. Protein composition of the dystrophic muscle, *Bull. Jpn. Soc. Sci. Fish.*, 36, 1231, 1970.
36. **Hashimoto, Y., Okaichi, T., Watanabe, T., and Furukawa, A.,** Muscular dystrophy of carp due to oxidized oil and the preventive effect of vitamin E, *Bull. Jpn. Soc. Sci. Fish.*, 32, 64, 1966.

37. **Watanabe, T., Tsuchiya, T., and Hashimoto, Y.,** Effect of DPPD and ethoxyquin of the muscular dystrophy of carp induced by oxidized saury oil, *Bull. Jpn. Soc. Sci. Fish.*, 33, 843, 1967.

38. **Watanabe, T. and Hashimoto, Y.,** Toxic components of oxidized saury oil inducing muscular dystrophy in carp, *Bull. Jpn. Soc. Sci. Fish.*, 34, 1131, 1968.

39. **Watanabe, T., Takeuchi, T., and Wada, M.,** Dietary lipid levels and α-tocopherol requirement of carp, *Bull. Jpn. Soc. Sci. Fish.*, 47, 1585, 1981.

40. **Watanabe, T., Matsuura, Y., and Hashimoto, Y.,** Effect of natural and synthetic antioxidants on the incidence of muscle dystrophy of carp induced by oxidized saury oil, *Bull. Jpn. Soc. Sci. Fish.*, 32, 887, 1966.

41. **Murai, T. and Andrews, J. W.,** Interactions of dietary α-tocopherol, oxidized menhaden oil and ethoxyquin on channel catfish (*Ictalurus punctatus*), *J. Nutr.*, 104, 1416, 1974.

42. **Takeuchi, M.,** Effect of vitamin E and ethoxyquin on absorption of lipid hydroperoxide in carp, *Bull. Jpn. Soc. Sci. Fish.*, 38, 155, 1972.

43. **Aoe, H., Abe, I., Saito, T., Fukawa, H., and Koyama, H.,** Preventive effects of tocopherols on muscular dystrophy of young carp, *Bull. Jpn. Soc. Sci. Fish.*, 38, 845, 1972.

44. **Higashi, H., Terada, K., Morinage, H., and Nakahira, T.,** Studies on roles of tocopherols in fish. II. Mutual conversion of tocopherol-homologues in carps, when α, γ or δ tocopherol was given to them (preliminary report), *Vitamins*, 45, 121, 1972.

45. **Furuichi, M. and Yone, Y.,** Effect of dietary dextrin levels on the growth and feed efficiency, the chemical composition of liver and dorsal muscle, and the absorption of dietary protein and dextrin in fishes, *Bull. Jpn. Soc. Sci. Fish.*, 46, 225, 1980.

46. **Furuichi, M. and Yone, Y.,** Availability of carbohydrate in carp and red sea bream, *Bull. Jpn. Soc. Sci. Fish.*, 48, 945, 1982.

47. **Furuichi, M., Shitanda, K., and Yone, Y.,** Studies on nutrition of red sea bream-V. Appropriate supply of dietary carbohydrate, *Rep. Fish. Res. Lab., Kyushu Univ.*, No. 1, 91, 1971.

48. **Murai, T., Akiyama, T., and Nose, T.,** Effects of glucose chain length of various carbohydrates and frequency of feeding on their utilization by fingerling carp, *Bull. Jpn. Soc. Sci. Fish.*, 49, 1607, 1983.

49. **Ogino, C., Chiou, J.-Y., and Takeuchi, T.,** Protein nutrition in fish-VI. Effects of dietary energy sources on the utilization of proteins by rainbow trout and carp, *Bull. Jpn. Soc. Sci. Fish.*, 42, 213, 1976.

50. **Aoe, H., Masuda, I., Mimura, T., Saito, T., Komo, A., and Kitamura, S.,** Water soluble vitamin requirements for carp-VI. Requirement for thiamin and effects of antithiamins, *Bull. Jpn. Soc. Sci. Fish.*, 35, 459, 1969.

51. **Ogino, C.,** B vitamin requirements of carp, *Cyprinus carpio*. II. Requirements for riboflavin and pantothenic acid, *Bull. Jpn. Soc. Sci. Fish.*, 33, 351, 1967.

52. **Takeuchi, T., Takeuchi, L., and Ogino, C.,** Riboflavin requirement in carp and rainbow trout, *Bull. Jpn. Soc. Sci. Fish.*, 46, 733, 1980.

53. **Ogino, C.,** B vitamin requirements of carp, *Cyprinus carpio*. I. Deficiency symptoms and requirements of vitamin B_6, *Bull. Jpn. Soc. Sci. Fish.*, 31, 546, 1965.

54. **Aoe, H., Masuda, I., and Takada, T.,** Water-soluble vitamin requirements of carp. III. Requirement for niacin, *Bull. Jpn. Soc. Sci. Fish.*, 33, 681, 1967.

55. **Ogino, C., Watanabe, T., Kakino, J., Iwanaga, N., and Mizuno, M.,** B vitamin requirements of carp. III. Requirement for biotin, *Bull. Jpn. Soc. Sci. Fish.*, 36, 734, 1970.

56. **Ogino, C., Uki, N., Watanabe, T., Iida, A., and Ando, K.,** B vitamin requirements of carp. IV. Requirement for choline, *Bull. Jpn. Soc. Sci. Fish.*, 36, 1140, 1970.

57. **Aoe, H. and Masuda, I.,** Water-soluble vitamin requirements of carp. II. Requirements for *p*-aminobenzoic acid and inositol, *Bull. Jpn. Soc. Sci. Fish.*, 33, 674, 1967.

58. **Aoe, H., Masuda, I., Mimura, T., Saito, T., and Komo, A.,** Requirement of young carp for vitamin A, *Bull. Jpn. Soc. Sci. Fish.*, 34, 959, 1968.

59. **Dabrowski, K., Hinterleitner, S., Sturmbauer, C., El-Fiky, N., and Wieser, W.,** Do carp larvae require vitamin C? *Aquaculture*, 72, 295, 1988.

60. **Ogino, C. and Takeda, H.,** Mineral requirements in fish-III. Calcium and phosphorus requirements in carp, *Bull. Jpn. Soc. Sci. Fish.*, 42, 793, 1976.

61. **Ogino, C., Takeuchi, L., Takeda, H., and Watanabe, T.,** Availability of dietary phosphorus in carp and rainbow trout, *Bull. Jpn. Soc. Sci. Fish.*, 45, 1527, 1979.

62. **Ogino, C. and Chiou, J.-Y.,** Mineral requirements in fish-II. Magnesium requirement of carp, *Bull. Jpn. Soc. Sci. Fish.*, 42, 71, 1976.

63. **Ogino, C. and Yang, G.-Y.,** Requirement of carp for dietary zinc, *Bull. Jpn. Soc. Sci. Fish.*, 45, 967, 1979.

64. **Sakamoto, S. and Yone, Y.,** Iron deficiency symptoms of carp, *Bull. Jpn. Soc. Sci. Fish.*, 44, 1157, 1978.

65. **Ogino, C. and Yang, G.-Y.,** Requirement of carp and rainbow trout for dietary manganese and copper, *Bull. Jpn. Soc. Sci. Fish.*, 46, 455, 1980.

66. **Yone, Y. and Toshima, N.,** The utilization of phosphorus in fish meal by carp and black sea bream, *Bull. Jpn. Soc. Sci. Fish.*, 45, 753, 1979.

67. **Takamatsu, C., Endoh, E., Hasegawa, T., and Suzuki, T.,** Effect of phosphate supplemented diet on growth of carp, *Suisanzoshoku*, 23, 55, 1975.

68. **Shitanda, K., Wagatsuma, R., and Ukita, M.,** Effect of phosphorus supplement to commercial diet on growth, feed efficiency, chemical component of serum and body with carp, *Suisanzoshoku*, 27, 26, 1979.

69. **Satoh, S., Yamamoto, H., Takeuchi, T., and Watanabe, T.,** Effects on growth and mineral composition of carp of deletion of trace elements or magnesium from fish meal diet, *Bull. Jpn. Soc. Sci. Fish.*, 49, 431, 1983.

70. **Satoh, S., Takeuchi, T., and Watanabe, T.,** Availability to carp of manganese in white fish meal and of various manganese compounds, *Nippon Suisan Gakkaishi*, 53, 825, 1987.

71. **Satoh, S., Izume, K., Takeuchi, T., and Watanabe, T.,** Availability to carp of manganese contained in various types of fish meals, *Nippon Suisan Gakkaishi*, 55, 313, 1989.

72. **Watanabe, T., Takeuchi, T., Satoh, S., Wang, K.-W., Ida, T., Yaguchi, M., Nakada, M., Amano, T., Yoshijima, S., and Aoe, H.,** Development of practical carp diets for reduction of total nitrogen loading on water environment, *Nippon Suisan Gakkaishi*, 53, 2217, 1989.

73. **Kariya, T.,** On the feeding habit of fish, *Suisanzoshoku*, 4, 1, 1956.

74. **Tominaga, M.,** Feeding of common carp culture, *Yoshoku*, 1, 52, 1964.

75. **Kurihara, N.,** Amount of food given to common carp reared in net fish cage placed in lake, *Suisanzoshoku*, 13, 197, 1966.

76. **Tazaki, S.,** Studies of feeding to one year common carp, *Yoshoku*, 15, 57, 1978.

EEL, *ANGUILLA* Spp.

Shigeru Arai

INTRODUCTION

Eels are generally classified as a warmwater fish, and 19 species, including subspecies of the *Anguilla* genus, are distributed throughout the world. The following four species, *Anguilla japonica* in the Far East, *Anguilla anguilla* in Europe, *Anguilla rostrata* in North America, and *Anguilla australis* in Australia and New Zealand are commercially important. Extensive eel culture of the European eel *A. anguilla* has been conducted for many years in lagoons of the Mediterranean coast, however, intensive eel culture has been developed mostly in the Far East, especially in Japan using the Japanese eel *A. japonica*. Japanese eel has been cultured in Japan since 1880 and the annual production was 36,994 tons in 1987.[1] Eel culture in Taiwan was started about 20 years ago and the total production is estimated to be about 35,000 tons yearly. Most of eels produced in Taiwan are exported to Japan. Commercial intensive eel culture operations are not well established in Europe, North America, and Australasia with most of the production coming from catches in natural waters. Total production of eels in the world was 92,061 tons in 1987.[1] Production of cultured eel has increased, whereas the commercial catch of eels in natural waters in the world has decreased yearly. Since the demand for eel is still increasing in the world, production of eels by culture will be more important in the future.

LABORATORY CULTURE

Eels breed in the sea after migrating long distances from rivers where they lived. Elvers of the European eel enter rivers 3 years after hatching and of the Japanese eel after 1 year. No one has been successful in obtaining elvers by artificial breeding, therefore elvers must be obtained from the natural waters. Elvers caught in natural waters have to be trained to eat experimental diets. Though live tubificid (family Tubificidae) worms, fresh fish, and short-necked clams have been used as starter feeds for many years, tubificid worms are the most desirable food for elvers. After stocking in the nursery tank, the elvers are fed chopped tubificid worms for the first 2 to 3 d. In about 1 week, the elvers may be trained to feed on the worms. Beef liver can be used as a substitute of the tubificid worms. After the hard connective tissue is removed, the liver is finely ground with a meat mincer and mixed thoroughly. If the ground liver is placed in an airtight container and kept in a freezer at below −20°C, it can be used for a longer period. Recently, packed and frozen, prepared feeds for elver have been developed as a substitute of live tubificid worms. These are being used by many farmers in Japan, because of the shortage of tubificid worms during the winter season and because they are convenient to use.

After the eels are trained to the starter feed, the feed is changed gradually from the starter to a formulated eel diet or an experimental purified diet by mixing of the two diets over several days. Examples of experimental diets that have been used for eels are shown in Table 1. After acclimation to the diet, the elvers are fed only the amount that they will consume in a short time. During these training days, the elvers are fed the diet two to four times daily. When the elvers become larger and all the eels are feeding properly, then the feeding frequency may be decreased to two times daily. In commercial pond eel culture, it is customary to feed only one time in the morning, because the oxygen content of the water is highest during the day time due to photosynthesis of algae in the pond.

The daily feeding rate of a formulated feed on a dry basis at a water temperature of 25°C

TABLE 1
Examples of Purified and Semipurified Diets for Eel

| Ingredient | Diet no. and reference | | | | |
	1[2]	2[3]	3[4]	4[5]	5[6,7]
Casein	54.0	49.0	—	—	49.0
Gelatin	15.0	—	—	—	—
Amino acid mix	1.5[a]	1.0[b]	46.7[c]	—	1.0[b]
White fish meal	—	—	—	70.0	—
Pollack liver oil	3.0	2.7	2.7	1.0	—
Corn oil	6.0	5.3	5.3	2.0	—
Methyl laurate	—	—	—	—	6.0
Methyl linoleate	—	—	—	—	0.5
Methyl linolenate	—	—	—	—	0.5
Dextrin	8.0	21.0	14.0	5.0	21.0
Cellulose	6.0	4.0	14.6	6.0	0.5
CMC	—	10.0	10.0	10.0	10.0
Vitamin mix[8]	6.0	4.0	14.6	4.0	4.5
Mineral mix[8]	7.0	7.0	6.7	2.0	7.0
Water	200	180	100	150	180

[a] Supplied 0.5 g tryptophan and 1.0 g cystine.
[b] Supplied 0.7 g arginine and 0.3 g cystine.
[c] Supplied 2.67 g arginine·HCl; 1.33 g histidine·HCl; 2.67 g isoleucine; 4.00 g leucine; 3.34 g lysine·HCl; 1.33 g methionine; 2.33 g phenylalanine; 2.00 g threonine; 0.67 g tryptophan; 3.00 g valine; 3.00 g alanine; 5.00 g aspartic acid; 0.67 g cystine; 6.67 g glutamic acid; 3.34 g glycine; 2.67 g proline; and 2.00 g tyrosine.

is 6 to 8% of body weight for elvers and small eels, and 2 to 3% for larger eels. For feeding raw fish flesh, the feeding rate on a wet weight basis at 25°C is 20 to 30% of body weight for elvers and small eels, and about 10% for larger eels. Since these are standard feeding rates, changes in the daily feeding rate should be made based on the observed feeding activity of the eels and on water quality conditions.

Arai et al.[2] have developed an aquarium design for feeding experiments for eels. The tanks were made of polyvinyl chloride and a small one had dimensions of $20 \times 20 \times 50$ cm. A narrow rim at the top of the tank prevented elvers from escaping and nets covered the inlet, outlet, and overflow points. For larger tanks, two to three outlet and overflow points should be provided, because the net mesh is often covered with mold due to the rich nutrients in the feed and the high water temperature. Suitable water temperatures for optimum growth of Japanese eel and European eel is 23 to 30°C and 20 to 23°C, respectively. Water temperature should be controlled at about 25°C for Japanese eels and at about 22°C for European eels and supplied at a rate of 0.3 to 0.6 l/min to the tanks for optimum water conditions. The tank water must be 12 to 15 cm deep and be aerated by air stones for supplying oxygen. Each tank should be covered with a polyvinyl chloride plate.

Thirty to fifty elvers can be put in an aquarium of the size described above. About 1 to 3 g elvers are the most suitable size to use for feeding experiments. The elvers are fed twice daily because the feed intake must be increased at these high water-temperature conditions. They should also be fed at least once a day during weekends in order to maintain their growth rate and to keep them from becoming cannibalistic caused by starvation. A paste diet using carboxymethyl cellulose (CMC) as a binder (diet 2 to 5 in Table 1) can be extruded like a noodle into the tank from a 50-ml nylon syringe. The size of the diameter of the noodle must be changed with the growth of the eels. The tanks should be cleaned each day with a nylon brush. The fish are weighed individually at 2-week intervals by anesthetizing with a 1.2% urethane solution or other anesthetic.

It is well known that eels show considerable variation in growth rates in experimental tanks as well as in cultural ponds. Due to this characteristic, farmers continually sort the eels during culture. In order to succeed in conducting feeding experiments with eels, it is desirable to eliminate groups of eels that have exhibited slow growth rates in the nursery tank or during the early stages of feeding.

Two types of water systems are used in practical eel culture and also for laboratory feeding studies. The flow-through water system has the advantage of being able to control the experimental conditions in the tank or pond, however, an adequate water supply is needed. On the other hand, a still-water system will not need as much water, but conditioning of the water in a pond is difficult. Still-water systems are not suitable for experimental nutritional studies because the water can become polluted with nutrients that may affect the fish. Well water is usually used, but river water can be used if care to avoid water pollution is taken. Generally, the oxygen content of well water is lower than required, thus the water should be exposed to the air for a while before use and should be aerated with aeration devices. Eels cannot breathe and come up to the surface when the oxygen content of the water falls below 1 mg/l.

NUTRIENT REQUIREMENTS

PROTEIN AND AMINO ACIDS

Nose and Arai[3] determined the dietary protein requirement of Japanese eel using a purified casein diet (diet 2 in Table 1). Eels fed a nonprotein diet showed a marked decrease in body weight, and a slight decrease in body weight was observed in eels fed a 8.9% protein diet. Eels fed diets containing more than 13.4% protein attained a positive weight gain, and the best growth was observed in eels fed diets containing more than 44.5% protein. Young Japanese eels require about 45% dietary protein for normal growth.

Arai et al.[4] determined the qualitative amino acid requirements for two species of eel, *A. anguilla* and *A. japonica*, using an amino acid test diet (diet 3 in Table 1). Eels fed diets deficient in each of alanine, aspartic acid, cystine, glutamic acid, glycine, proline, and tyrosine grew as well as those fed the complete amino acid diet. The eels fed diets deficient in each of arginine, histidine, isoleucine, leucine, lysine, methionine, phenylalanine, threonine, tryptophan, and valine failed to grow until the deleted amino acid was added to the ration. Thus, the eels required the same 10 amino acids that have been reported to be essential for growth of other fish. It may be noteworthy that eels seem to be much more sensitive to the lack of essential amino acids. A loss of appetite was recognized in a period as short as 3 d, and recovery of appetite in the deficient eels was also rapid, when the deleted essential amino acid was added to the ration. The effects of supplemental L-tryptophan and DL-methionine or L-cystine on the growth of the eel were studied by Arai et al.[2] The growth rate was doubled by replacing DL-methionine with an equal amount of L-cystine in the purified diet. Cystine is a nonessential amino acid.[4] However, it seems to have some growth-promoting effect in eels, and these results suggest the importance of proper amino acid balance in the protein nutrition of the eel.

Growth studies were conducted with young Japanese eels to determine the minimal quantitative requirement for each essential amino acid. A mixture of crystalline amino acids similar in composition to beef liver protein was used as the nitrogen source and its level was fixed at 46.7% (crude protein = 37.7%) in the basal diet. For each essential amino acid, nine different dietary levels were prepared by replacing the test amino acid with an equal amount of L-alanine. The diets were adjusted to pH 6.2 to 6.5 with 25% NaOH. Feeding experiments were conducted for 4 to 8 weeks for each amino acid.

The quantitative essential amino acid requirements of eel expressed as a percentage of dietary protein are as follows: 4.2% of arginine; 2.1% of histidine; 4.1% of isoleucine; 5.4% of leucine; 5.3% of lysine; 3.2% of methionine (cystine = 0) and 2.4% (cystine = 2.7%); 5.6% of phenylalanine (tyrosine = 0) and 3.2% (tyrosine = 5.3%); 4.1% of threonine; 1.0% of tryptophan; and 4.1% of valine.[9]

LIPIDS

Arai et al.[2] examined the effects of supplemental lipids on the growth of young Japanese eel using a purified casein-gelatin diet. As lipid sources, a mixed oil consisting of corn oil and cod liver oil in a 2:1 ratio was the most favorable, corn oil the second, and cod liver oil the least. Eels fed diets containing the higher lipid level deposited a larger amount of fat in their body cavity.

Takeuchi et al.[7] determined the essential fatty acid (EFA) requirement of the Japanese eel using various diets containing methyl esters of both 18:2n-6 and 18:3n-3 in different ratios as the lipid source (diet 5 in Table 1). In addition, the effect of n-3 HUFAs (a mixture of 20:5n-3 and 22:6n-3) on the growth of the eel was compared with that of 18:2n-6 and 18:3n-3. Eels fed a fat-free or an EFA-deficient diet had reduced growth rates and low feed efficiency. The addition of either 18:2n-6 or 18:3n-3 to the diets improved growth rates, 18:3n-3 being more effective. The best weight gain was obtained in the fish receiving a diet containing 0.5% 18:2n-6 and 0.5% 18:3n-3. Eels fed the latter diet had growth rates that were almost comparable to that of fish fed the control diet containing a mixture of corn oil and cod liver oil. The addition of n-3 fatty acids, not only 18:3n-3 but also n-3 HUFAs, to the EFA-deficient diet improved the growth rate and feed efficiency. The supplemental effect of 1% n-3 HUFA was almost the same as that of 1% 18:3n-3. Thus, the dietary requirement of Japanese eel for 18:2n-6 and 18:3n-3 is estimated to be about 0.5% of each or 1% for 18:3n-3.

CARBOHYDRATES

Although commercial eel feeds in Japan contain high levels of alpha-potato starch as a carbohydrate source and as a binder (diet 1 in Table 2), little information is available on carbohydrate nutrition of the eel.

Takii et al.[10] reported a 77.1 to 87.8% apparent digestibility of alpha-potato starch and an 85.3 to 88.4% of protein in 78 g Japanese eels.

VITAMINS

Hashimoto et al.[11] reported the qualitative thiamine requirement and deficiency signs of Japanese eel using short-necked clam diet. Then, a purified vitamin test diet (diet 1 in Table 1) for Japanese eel was formulated[2] by modifying the vitamin test diet developed for chinook salmon.[8] With this diet, Arai et al.[12] reported the qualitative requirements of young Japanese eel for the water-soluble vitamins and their deficiency signs. Young Japanese eels require 11 water-soluble vitamins but not *p*-aminobenzoic acid. The deficiency signs of these vitamins are summarized in Table 3.

In the experiments, the mortality of eels fed diets devoid of each vitamin was low during the experimental period, except for pantothenic acid deficiency. Most of the deficiency signs disappeared soon after feeding the missing vitamin. The deficiency signs observed in eels fed ascorbic acid-deficient diets were very similar to scurvy found in warm-blooded animals. Examination of eels fed the choline or the inositol-deficient diet revealed a similar discoloration of the intestine. Both vitamins are components of phospholipids and seem to play a similar role in lipid metabolism. Elimination of each of riboflavin, pantothenic acid, and niacin from the diet induced a dermatitis, including hemorrhage and congestion in the fins. The skin lesions induced by the lack of these water-soluble vitamins may be the cause of the fungal disease prevalent on eel farms.

TABLE 2
Examples of Practical Diets for Eel

Ingredient	Diet no. and reference		
	1[16]	2[13,14]	3[10]
Casein	—	—	30.0
White fish meal	65.0	71.0	40.0
Pollack liver oil	—	—	5.0
Alpha potato starch	22.5	20.0	11.0
Soybean flour	5.0	—	—
Yeast	3.0	—	—
Liver meal	2.0	—	—
Wheat gluten	—	—	8.0
Carboxymethyl cellulose	—	3.0	—
Binder	—	—	2.0
Vitamin mix	0.5	1.0	2.0
Mineral mix	2.0	5.0	2.0
Water	100	150	160

TABLE 3
Vitamin Deficiency Signs in Eels[11,12]

Thiamine	Trunk-winding, ataxia, hemorrhage and congestion in fins, dark coloration, and sluggish movement
Riboflavin	Hemorrhage and congestion in fins, dermatitis, photophobia, and sluggish movement
Pyridoxine	Nervous disorders, epileptiform fits, and convulsions
Pantothenic acid	Abnormal swimming, ataxia, mortality, hemorrhage in epidermis, and dermatitis
Niacin	Abnormal swimming, ataxia, anemia, hemorrhage in epidermis, and dermatitis
Biotin	Abnormal swimming and dark coloration
Folic acid	Dark coloration and poor growth
Vitamin B_{12}	Poor growth
Choline	White-grey intestine
Inositol	White-grey intestine
Ascorbic acid	Hemorrhage in fins, head and skin, and lower jaw erosion
p-aminobenzoic acid	None
Vitamin E_6	Hemorrhage in fins and skin, and dermatitis

Yamakawa et al.[6] reported the vitamin E deficiency signs and quantitative requirement of α-tocopherol using vitamin-free casein as the main protein source and fatty acid methyl esters (diet 5 in Table 1) as lipid sources. The experimental fish were fed a diet devoid of α-tocopherol for 4 weeks before the feeding experiment started. In the α-tocopherol deficient group, poor appetite, poor growth, hemorrhage and congestion in the fins, and dermatitis on the skin were observed as vitamin E deficiency signs. From the results of the feeding experiment, the minimal requirement of α-tocopherol for growth of young eels was estimated to be about 20 mg in 100 g dry diet. Other fat-soluble vitamin requirements of eels have not been established.

MINERALS

It is well known that fish can absorb minerals from the water. However, Arai et al.[2] found that the addition of a dietary mineral mixture to the purified casein-gelatin diet (diet 1 in Table 1) was necessary for proper growth of young Japanese eel. The eels fed a diet devoid of a mineral mixture or a diet with the mineral mix added at the 1% level showed a cessation of growth in 2 weeks, followed by a gradual loss of body weight and high mortality. Eels fed diets containing higher levels of the mineral mixture showed better growth. The optimum level of the mineral mixture needed in the purified casein-gelatin diet for growth of eels in a flow-through water culture system was estimated to be about 8%, which was twice the level recommended in the purified test diets for salmonids. These results obviously indicate that the eel requires a higher level of supplementation of minerals in their diet than salmonids.

Because the Japanese eel required a high level of mineral mixture in the purified casein-gelatin diet for optimal growth, the need for mineral mixture supplementation to fish meal diets for young Japanese eel was examined.[5] The best growth was observed in fish fed a diet containing 2% of the mineral supplement. The growth rate was twice that of eels fed diet devoid of the mineral mixture. Eels fed the diet devoid of the mineral mixture had hypochromic microcytic anemia, whereas no abnormalities were observed on the blood of eels fed diets supplemented with the mineral mixture. The hematological observations indicated that the minerals in the fish meal seemed to be insufficient, qualitatively or quantitatively, to maintain proper health of eels under these experimental conditions. Although fish meal contains a high amount of crude ash, it either lacks some essential minerals or has certain mineral constituents that are not readily utilized by the eel. Recently Park and Shimizu[13,14] reported that the aluminum and zinc requirements of Japanese eel are 1.5 mg and 5 to 10 mg/100 g diet, respectively, using a white fish meal diet (diet 2 in Table 2).

The requirements of calcium, magnesium, and iron for growth of eel have been studied using a purified casein diet (diet 2 in Table 1), and the phosphorus requirement was studied using an amino acid test diet (diet 3 in Table 1).[4] Eels fed calcium- or phosphorus-deficient diets gradually lost their appetite in a week, followed by reduced growth. Eels fed magnesium- or iron-deficient diets showed poor appetite after 3 or 4 weeks. Eels fed the iron-deficient diet had hypochromic microcytic anemia at the end of feeding experiment. The minimal requirements of calcium, magnesium, phosphorus, and iron for young eels were estimated to be about 270, 40, 250 to 320, and 17 mg in a 100-g dry diet, respectively.[15]

PRACTICAL DIET FORMULATION

The composition of a typical eel diet used in Japan is shown in Table 2 (diet 1).[16] To 100 parts of the mash type diet, 5 to 10 parts of oil are added and mixed well, then 80 to 100 parts of water are added and the mixture is blended vigorously for a few minutes in a mixer just before feeding. The consistency of the diet is attributable to the presence of the alpha-potato starch, which will weaken quickly after preparation. Therefore, the diet should be fed immediately after preparation. The diet contains 22.5% (22 to 25%) of alpha-starch as a binder, however, other binders, such as alginic acid or CMC, may be substituted for the alpha-starch. In order to avoid the breakdown of the diet, 5 to 8% of CMC, 2 to 3% of alginic acid, and a small amount of some other binding material is necessary. The alpha-potato starch, which is one of the most expensive ingredients in commercial eel diets, can be replaced with a cheaper, low-grade wheat flour as a carbohydrate source.

SPECIAL DIETS

A pelleted dry diet can also be used for eels. However, eels are not skillful in taking pelleted diets like salmonids or other fishes. In order to use a pelleted dry diet, eels should be trained from the elver stage and sorting is needed more often than when the usual moist diets are fed. A pellet-type diet can be used for eel culture if proper sorting and selection of the proper pellet particle sizes are used at each growth stage. Otherwise, it is very difficult to obtain good results by feeding pelleted diets in practical eel culture.

As indicated above, considerable information is lacking for the nutrition of eels. It is necessary to continue studying the nutritional requirements of eels, not only *A. japonica* but also *A. anguilla* or *A. rostrata*, for further development of eel culture. Energy requirement studies are especially needed for the development of least-cost feed formulation. Formulations of the experimental diets shown in Table 1 can be used directly or indirectly for many types of nutritional studies.

Eel culture in Japan has developed rapidly after the development of formulated diets containing high amounts of good-quality fish meal. However, fish meal production has decreased yearly, whereas the demand for fish meal continues to increase. Now substitutes for fish meal are needed for the eel culture industry, otherwise it will suffer from a shortage and higher prices of the fish meal.

Additional research is also needed on the life history and reproductive behavior of eels. Due to the fluctuation in elver supply, eel culture operations have not been able to develop into a sound industry. For these reasons, studies on artificial spawning have been carried out in Japan and other countries. Hormone injection techniques have been used to assist the acceleration of maturation and spawning. However, artificial larval production has not been successful to date. Eels go for long periods such as 10 to 20 weeks without feeding prior to spawning, therefore studies on the nutrition and diet development for broodstock are needed.

REFERENCES

1. **FAO Yearbook,** Fishery Statistics, Catches and Landings, 64, 148, 1987.
2. **Arai, S., Nose, T., and Hashimoto, Y.,** A purified test diet for the eel, *Anguilla japonica, Bull. Freshwater Fish. Res. Lab.*, 21, 161, 1971.
3. **Nose, T. and Arai, S.,** Optimum level of protein in purified diet for eel, *Anguilla japonica, Bull. Freshwater Fish. Res. Lab.*, 22, 145, 1972.
4. **Arai, S., Nose, T., and Hashimoto, Y.,** Amino acids essential for the growth of eels, *Anguilla anguilla* and *A. japonica, Bull. Jpn. Soc. Sci. Fish.*, 38, 753, 1972.
5. **Arai, S., Nose, T., and Kawatsu, H.,** Effect of minerals supplemented to the fish meal diet on growth of eel, *Anguilla japonica, Bull. Freshwater Fish. Res. Lab.*, 24, 95, 1974.
6. **Yamakawa, T., Arai, S., Shimma, Y., and Watanabe, T.,** Vitamin E requirement for Japanese eel, *Vitamins*, 49, 62, 1975.
7. **Takeuchi, T., Arai, S., Watanabe, T., and Shimma, Y.,** Requirement of eel *Anguilla japonica* for essential fatty acids, *Bull. Jpn. Soc. Sci. Fish.*, 46, 345, 1980.
8. **Halver, J. E.,** Nutrition of salmonoid fishes. III. Water-soluble vitamin requirements of chinook salmon, *J. Nutr.*, 62, 225, 1957.
9. **Lovell, T.,** *Nutrition and Feeding of Fish*, Van Nostrand Reinhold, New York, 1989, 26.
10. **Takii, K., Shimeno, S., Takeda, M., and Kamekawa, S.,** The effect of feeding stimulants in diet on digestive enzyme activities of eel, *Bull. Jpn. Soc. Sci. Fish.*, 52, 1449, 1986.
11. **Hashimoto, Y., Arai, S., and Nose, T.,** Thiamine deficiency symptoms experimentally induced in the eel, *Bull. Jpn. Soc. Sci. Fish.*, 36, 791, 1970.
12. **Arai, S., Nose, T., and Hashimoto, Y.,** Qualitative requirements of young eels *Anguilla japonica* for water-soluble vitamins and their deficiency symptoms, *Bull. Freshwater Fish. Res. Lab.*, 22, 69, 1972.
13. **Park, C. W. and Shimizu, C.,** Quantitative requirements of aluminum and iron in the formulated diets and its interrelation with other minerals in young eel, *Nippon Suisan Gakkaishi*, 55, 111, 1989.
14. **Park, C. W. and Shimizu, C.,** Suitable level of zinc supplementation to the formulated diets in young eel, *Nippon Suisan Gakkaishi*, 55, 2137, 1989.
15. **Nose, T. and Arai, S.,** Recent advances in studies on mineral nutrition of fish in Japan, in *Advances in Aquaculture*, Pillay, T. V. R. and Dill, W. A., Eds., Fishing News, Farnam, England, 1979, 584.
16. **Lovell, T.,** *Nutrition and Feeding of Fish*, Van Nostrand Reinhold, New York, 1989, 227.

FLATFISH, TURBOT, SOLE, AND PLAICE

Jean Guillaume, Marie-Francoise Coustans, Robert Métailler,
Jeannine Person-Le Ruyet, and Jean Robin

INTRODUCTION

Flatfish belong to the order of Heterosomata; five species belonging to three families have been or are currently being evaluated for culture: plaice, *Pleuronectes platessa*; sole, *Solea solea*; and turbot, *Scophthalmus maximus* or *Psetta maxima*; and more recently halibut, *Hippoglossus hippoglossus*; and the Japanese flounder (hirame), *Paralychthis olivaceus*. The success of flatfish culture has been quite variable, i.e., the rearing of plaice has been discontinued, that of sole is very limited, while turbot culture is presently developing in Europe. Halibut culture seems promising in Scandinavia, but it is still just being developed, while hirame culture is expanding very rapidly in Japan. Unfortunately, nutritional experiments concerning the two latter-mentioned species are very scarce and quantitative nutritional recommendations are still lacking. Therefore most of the information presented in this review will be based on data for turbot, sole, and plaice.

NUTRIENT REQUIREMENTS

PROTEIN AND AMINO ACIDS

The requirement of young turbot (10 g) has been studied using semipurified diets composed of casein, gelatin, amino acids, oil, and dextrin, with crude protein content varying from 32 to 59.5%.[1] The results indicated the higher the protein level, the higher the growth rate. Therefore, with such high-energy diets (roughly 17.5 MJ/kg or 4.2 Mcal/kg), the protein requirement exceeded 60% of the diets. Unfortunately, no data are available on the requirement of turbot measured with practical diets or with larger fish.

In sole, studies on the protein requirement were made using semipurified diets based on casein and gelatin with protein levels ranging from 24 to 77%.[2] The best performance was obtained with 57 to 58% protein, irrespective of the lipid level. In these experiments, detrimental effects of excessive protein levels were apparent.

Two studies have been conducted with plaice. Very small fish (3 g) fed casein-based diets supplemented with amino acids had the best growth rate when fed diets containing the highest protein level (70%).[3] In a subsequent experiment, 10 to 20 g fish fed a more conventional diet containing cod muscle, shrimp meal, oil, and dextrin showed the best weight gain at 57% crude protein.[4]

These data appear to indicate that flatfish have a very high protein requirement, at least when expressed as a percentage of the diet. This result does not mean that the absolute requirement of flatfish (expressed as grams of protein per day per fish of a given weight) is higher than that of other fish, since protein efficiency expressed as protein efficiency ratios and protein utilization coefficients were high in all of the experiments cited.

The protein requirement of any animal depends, among other factors, on quality of the dietary protein and more precisely on its biological value, i.e., on the balance of essential amino acids. In both plaice and sole the essentiality of amino acids has been determined by measuring the incorporation of labeled carbon from glucose into amino acids.[5] The results clearly demonstrate that plaice and sole require the same essential amino acids as other fish, i.e., arginine, methionine, valine, threonine, isoleucine, leucine, lysine, histidine, and phenylalanine. Although data are missing, tryptophan should be added to this list, whereas cystine and tyrosine should probably be considered as semiessential amino acids, being derived from methionine and phenylalanine, respectively. Proline was synthesized very slowly by plaice and sole. This slow rate could possibly indicate a requirement for proline similar to that of birds for glycine. The quantitative amino acid requirements have not been determined and the influence of the amino acid profile on fish performances has not been studied.

ENERGY

Very few data are available on the digestive or metabolizable energy of diets or feedstuffs for flatfish. The digestibility of different types of starch has been studied in turbot.[6] The results indicated good utilization of cooked starch, even when added at 20% of the diet, but not for raw corn starch.

Because of this lack of data, authors and feed manufacturers assume that flatfish digest the usual sources of nutrients with similar efficiency as other carnivorous fish. Such a hypothesis has to be considered as very provisional. In turbot, a high content of lipid was observed in feces of fish fed 10 or 15% crude lipid;[1] this fecal excretion, which indicated a decrease in digestibility, was more pronounced at high dietary protein levels (over 50%). These results differ from similar studies in salmonids.

On the other hand, several experiments have shown an unusually high ability for sole to digest and metabolize carbohydrates, partly due to strong amylase and maltase activities.[7,8]

In one of the few studies referring to energy in flatfish, dietary metabolizable energy (ME) was calculated using the following physiological fuel values: 4 kcal/g carbohydrate, 9.5 kcal/g lipid, and 4.5 kcal/g protein.[9] Despite the many possible critics, we have chosen to recalculate data obtained by others by this method to render the cited experiments more comparable. In turbot fed purified diets high in cellulose,[9] a constant improvement in performance was observed when ME levels increased from 7.8 to 13.2 MJ/kg (1.86 to 3.15 Mcal/kg). Protein utilization and feed efficiency values improved in a similar manner, indicating that the optimal energy level was not reached. In a study on slightly larger fish,[1] energy levels ranging from 15.6 to 19.2 MJ/kg (3.73 to 4.63 Mcal/kg) were tested and found to be influenced by the dietary protein content, i.e., at low protein levels (37.5%), growth rates improved with increases in dietary energy, while at higher protein levels (69.8%) the reverse was true. This apparent detrimental effect of high-energy diets may be due to a detrimental effect of lipid, which has been shown to be poorly digested in high-energy diets, thus making energy level estimations very inaccurate. It is impossible to deduce any clear optimal dietary energy level from these few studies. At present, levels between 13 and 16 MJ/kg appear to be adequate, but more information is needed in this area.

The best source of energy for flatfish is not known and probably cannot be simply defined. According to the hypothesis of Cowey et al.,[10] feeding diets containing a high lipid/carbohydrate ratio should favor nitrogen retention, i.e., exert the strongest protein-sparing effect, such ratios should be chosen at low protein content. On the other hand, feeding diets containing a high carbohydrate/lipid ratio should lead to better availability of nutrients for energy expenditure. This hypothesis is in very good agreement with the results obtained with turbot,[1] although no accurate estimation of the best lipid/carbohydrate can be obtained from the quoted works.

In plaice,[10] feeding diets containing carbohydrate (up to 10% dextrin plus 10% glucose) resulted in better growth and feed efficiency than feeding diets devoid of carbohydrate when compared at the same protein to energy ratio.

ESSENTIAL FATTY ACIDS

As in coldwater fish, fatty acids of the n-6 series are less efficient than those of the n-3 series in supporting growth of flatfish, indicating a main requirement for the latter series by these species. In addition, turbot was the first species of fish in which a low efficiency of short-chain polyunsaturated fatty acids (PUFA) was demonstrated.[11,12] The relative effects of feeding either C18 or long-chain n-3 fatty acids, such as 20:5n-3, eicosapentaenoic acid (EPA), and 22:6n-3, docosahexaenoic acid (DHA), were also determined using diets containing 45% protein and 10% lipid. The dietary essential fatty acid (EFA) requirement was estimated to be 0.8% of EPA + DHA.[13] In a subsequent experiment, the same authors demonstrated that a high level of linolenic acid resulted in a partial compensation of the lack of EPA and/or DHA and that 2.7% linolenic acid was needed to replace the 0.8% n-3 PUFA.[14] In other words, bioconversion of linolenic acid (elongation and desaturation) to EPA and DHA is not impossible in turbot, as first suggested,[11,12] but it occurs at a rate that is too slow for maximum growth.[15] No quantitative data on the requirement for n-6 PUFA is available, although it has been suggested that turbot require a correct balance between the three main PUFAs — 20:4n-6, 20:5n-3, and 22:6n-3 — in the diet.[16]

VITAMINS

The requirements for thiamine and pyridoxine have been reported for turbot. In both experiments, rather similar semipurified diets were fed to juveniles weighing from 15 to 20 g during periods of 12 to 16 weeks. The results indicated requirements of 0.6 to 2.6 mg thiamine/kg of diet[17] and 1.0 to 2.5 mg pyridoxine/kg of diet.[18] These values are clearly lower than the requirement values published for both coldwater and warmwater fish.[19,20] Although a quantitative requirement for ascorbic acid has not been determined, the effect of ascorbic acid deficiency has been studied.[20-22] Ascorbic acid-deficient turbot exhibit unusual deficiency signs, i.e., hypertyrosinemia leading to renal granulomatous nodules and finally to death. These deficiency signs were delayed by feeding excess levels of other vitamins.[23]

Other vitamins have not been studied in turbot. However, considering the vitamin nutrition data reported for salmonids[24] and red sea bream,[25] it seems prudent to suggest that other vitamins are needed by flatfish and to recommend the levels published for warmwater fish.[20]

MINERALS

Mineral requirement studies in flatfish are limited to only one report involving phosphorus and iron in sole.[26] Very small fish were fed casein-gelatin based diets supplemented with various amounts of monosodium phosphate. The best growth and phosphorus retention were found with 0.7% phosphorus. The same author tried to determine the requirement for iron using similar diets supplemented with iron citrate. No effect of iron supplementation on growth was observed, although erythrocyte number, hematocrit content, and hemoglobin concentration increased from 0 to 600 mg iron/kg diet. Current recommendations for warmwater fish should be used for the other minerals and other flatfish species.

FEEDING OF LARVAE, WEANED JUVENILES, AND BROODSTOCK

Turbot larvae are among the most difficult to rear. The main food used during the first month of life consists of live prey.[27] Because of this, the determination of nutritional requirements of larvae is very difficult. However, feeding rotifers with different n-3 PUFA enrichments to turbot larvae led to an estimated requirement for n-3 PUFAs of 1.3% of the dry matter of the rotifer.[28] Beneficial effects of trace mineral enrichment of rotifers fed as live food to turbot larvae have also been demonstrated, but no quantitative recommendation was published.[29]

During weaning some flatfish exhibit problems related to feeding behavior. Soft-texture pellet crumbles may allow good growth and survival, provided they are palatable. Research by Mackie and coworkers have led to the identification of feeding attractants specific for sole[30-32] and turbot.[32] These attractants mainly consist of nucleotides or nucleosides for turbot and of mixtures of amino acids, nucleotides, and nitrogenous bases for sole. By incorporating these attractants in weaning diets, some improvements have been observed in turbot, but mainly in sole. In turbot, dietary supplementation with inosine mainly improved survival rate.[33] In sole, the results were much more impressive, i.e., the chemical mixture was more efficient that extracts of prey and led to survival and specific growth rates amounting to 7 and 4.5 times, respectively, the corresponding values for the controls.[34,35] Recommendations were also given for the concentrations and duration of distribution.[34,35]

Broodstock are usually fed fresh food (trash fish and other marine animals) or a mixture of trash fish and undefined formulated food. Ascorbic acid injection has been shown to improve spawning results[36] but no feeding level was recommended.

FEED FORMULATIONS

Most of the experimental results discussed in this review should be treated as preliminary. In fact, certain experimental diets have led to impressive growth performance, such as those tested in turbot of 10 g by Caceres et al.,[1] where a feed conversion of 0.6 kg dry diet/kg wet gain was obtained, indicating that both the level and balance of nutrients were quite adequate for the fish. On the other hand, the amount of nutritional information is, without any doubt, insufficient for an efficient least-cost formulation. Due to the present lack of knowledge, several types of diets may be used, and experience will indicate the most appropriate (from a nutritional point of view) or most economical solutions.

In turbot, extremely high protein levels will probably be too expensive, and levels of 50 to 60% will perhaps be used. Higher protein levels could be more efficient, while levels around 50% will lead to a certain sparing of protein. This sparing effect could be enhanced by marine fish oil incorporation, at least when protein levels do not exceed 50%. Lipid levels exceeding 12% should be avoided when protein levels of 55% or higher are used. Fish meal can be replaced, at least in part, by other animal sources that may be less expensive, but data are not available on the potential use of plant protein sources.

Similar high-protein diets have to be used for sole, but less information exists concerning the appropriate lipid level. Since sole can generally digest and metabolize carbohydrates efficiently, they should be used by feed manufacturers as an energy source.

REFERENCES

1. **Caceres-Martinez, C., Cadena-Roa, M., and Métailler, R.,** Nutritional requirements of turbot (*Scophthalmus maximus*). I- A preliminary study of protein and lipid utilization, *J. World Maricult. Soc.*, 15, 191, 1984.
2. **Cadena-Roa, M.,** Etude expérimentale de l'alimentation de la sole (*Solea vulgaris* Q.) en élevage intensif, Thèse Doctorat de 3ème Cycle Univ. de Bretagne Occidentale, Brest, France, 1983.
3. **Cowey, C. B., Adron, J. W., Blair, A., and Pope, J.,** The growth of O-group plaice on artificial diets containing different levels of protein, *Helgoländer Wiss Meersunters.*, 20, 602, 1970.
4. **Cowey, C. B., Pope, J. A., Adron, J. W., and Blair, A.,** Studies on the nutrition of marine flatfish. The protein requirement of plaice (*Pleuronectes platessa*), *Br. J. Nutr.*, 28, 447, 1972.
5. **Cowey, C. B., Adron, J., and Blair, A.,** Studies on the nutrition of marine flatfish. The essential amino acid requirements of plaice and sole, *J. Mar. Biol. Ass. UK*, 50, 87, 1970.
6. **Jollivet, D., Gabaudan, J., and Métailler, R.,** Some effects of physical state and dietary level of starch, temperature and meal size on turbot (*Scophthalmus maximus* L.) digestive processes, *ICES CM* F:25, 1988.
7. **Clark, J., Naughton, J. E., and Stark, J. R.,** Metabolism in marine flatfish. I-Carbohydrate digestion in Dover sole (*Solea solea* L.), *Comp. Biochem. Physiol.*, 77B, 821, 1984.
8. **Cowey, C. B., Adron, J. W., Brown, D. A., and Shanks, A.,** Studies on the nutrition of marine flatfish. The metabolism of glucose by plaice (*Pleuronectes platessa*) and the effect of dietary energy source on protein utilization in plaice, *Br. J. Nutr.*, 33, 219, 1975.

9. **Adron, J. W., Blair, A., Cowey, C. B., and Shanks, A. M.,** Effects of dietary energy and dietary source on growth, feed conversion and body composition of turbot (*Scophthalmus maximus* L.), *Aquaculture*, 7, 125, 1976.

10. **Cowey, C. B., Brown, D. A., Adron, J. W., and Shanks, A. M.,** Studies on the nutrition of marine flatfish. The effect of dietary protein content on certain cell components and enzymes in the liver of *Pleuronectes platessa*, *Mar. Biol.*, 28, 207, 1974.

11. **Cowey, C. B., Adron, J. W., Owen, J. M., and Roberts, R. J.,** The effect of different dietary oils on tissue fatty acids and tissue pathology in turbot *Scophthalmus maximus*, *Comp. Biochem. Physiol.*, 53B, 399, 1976.

12. **Cowey, C. B., Owen, J. M., Adron, J. W., and Middleton, C.,** Studies on the nutrition of marine flatfish. The effect of different dietary fatty acids on the growth and fatty acid composition of turbot (*Scophthalmus maximus*), *Br. J. Nutr.*, 36, 479, 1976.

13. **Gatesoupe, F. J., Léger, C., Métailler, R., and Luquet, P.,** Alimentation lipidique du turbot (*Scophthalmus maximus* L.). I. Influence de la longueur de la chaîne des acides gras de la série ω3, *Ann. Hydrobiol.*, 8, 89, 1977.

14. **Gatesoupe, F. J., Léger, C., Boudon, M., Métailler, R., and Luquet, P.,** Alimentation lipidique du turbot (*Scophthalmus maximus* L.). II. Influence de la supplémentation en esters méthyliques de l'acide linolénique et de la complémentation en acides gras de la série ω9 sur la croissance, *Ann. Hydrobiol.*, 8, 247, 1977.

15. **Léger, C., Gatesoupe, F. J., Métailler, R., Luquet, P., and Frémont, L.,** Effect of dietary fatty acids differing by chain length and ω series on the growth and lipid composition of turbot *Scophthalmus maximus* L., *Comp. Biochem. Physiol.*, 64B, 345, 1979.

16. **Bell, M. V., Henderson, R. J., and Sargent, J. R.,** Changes in the fatty acid composition of phospholipids from turbot (*Scophthalmus maximus*) in relation to dietary polyunsaturated fatty acid deficiencies, *Comp. Biochem. Physiol.*, 81B, 193, 1985.

17. **Cowey, C. B., Adron, J. W., and Knox, D.,** Studies on the nutrition of marine flatfish. The thiamin requirement of turbot (*Scophthalmus maximus*), *Br. J. Nutr.*, 34, 383, 1975.

18. **Adron, J. W., Knox, D., Cowey, C. B., and Ball, G. T.,** Studies on the nutrition of marine flatfish. The pyridoxine requirement of turbot (*Scophthalmus maximus*), *Br. J. Nutr.*, 40, 261, 1978.

19. **National Research Council,** *Nutrient Requirements of Coldwater Fishes*, National Academy Press, Washington, D.C., 1981.

20. **National Research Council,** *Nutrient Requirements of Warmwater Fishes and Shellfishes*, National Academy Press, Washington, D.C., 1983.

21. **Tixerant, G., Aldrin, J. F., Baudin-Laurencin, F., and Messager, J. L.,** Syndrome granulomateux et perturbation du métabolisme de la tyrosine chez le turbot (*Scophthalmus maximus*), *Bull. Acad. Vet. de France*, 57, 75, 1984.

22. **Messager, J. L., Ansquer, D., Métailler, R., and Person-Le Ruyet, J.,** Induction expérimentale de l'hypertyrosinémie granulomateuse chez le turbot d'élevage (*Scophthalmus maximus*) par une alimentation carencée en acide ascorbique, *Ichtyophysiol. Acta*, 10, 201, 1986.

23. **Coustans, M. F., Guillaume, J., Métailler, R., and Dugornay, O.,** Effect of an ascorbic acid deficiency on tyrosinemia and renal granulomatous disease in turbot (*Pisces scophthalmidae*). Interaction with a slight polyhypovitaminosis, *Comp. Biochem. Physiol.*, 974, 145, 1990.

24. **Halver, J. E.,** *Fish Nutrition*, 2nd ed., Academic Press, San Diego, CA, 1988.

25. **Yone, Y.,** Nutritional studies of red sea bream, in *Proc. First Int. Conf. Aquaculture Nutr.*, Price, K. S., Shaw, W. N., and Danberg, D. S., Eds., University of Delaware, Lewes/Rehoboth, 1976, 39.

26. **Caceres-Martinez, C.,** Étude sur les besoins nutritionnels de la sole (*Solea vulgaris*) et du turbot (*Psetta maxima*), Thèse Doctorat de 3ème Cycle, Université de Bretagne Occidentale, Brest, France, 1984.

27. **Person-Le Ruyet, J., Baudin-Laurencin, F., Devauchelle, N., Métailler, R., Nicolas, J. L., Robin, J., and Guillaume, J.,** Culture of turbot (*Scophthalmus maximus*), in *Handbook of Mariculture, Vol. II: Finfish Aquaculture*, McVey, J. P., Ed., CRC Press, Boca Raton, FL, 1990, 21.

28. **Le Milinaire, C., Gatesoupe, F. J., and Stéphan, G.,** Approche du besoin quantitatif en acides gras longs polyinsaturés de la série n-3 chez le turbot (*Scophthalmus maximus*), *C.R. Acad. Sc. Paris*, 296, 917, 1982.

29. **Robin, H. J.,** The quality of living preys for fish larval culture: preliminary results on mineral supplementation, in *Aquaculture, A Biotechnology in Progress*, De Pauw, N., Jaspers, E., Ackefors, H., and Wilkins, N., Eds., European Aquaculture Society, Bredene, Belgium, 1989, 769.

30. **Mackie, A. M., Adron, J. W., and Grant, P.,** Chemical nature of feeding stimulants for the juvenile Dover sole *Solea solea* (L.), *J. Fish Biol.*, 16, 701, 1980.

31. **Mackie, A. M. and Mitchell, A.,** Further studies on the chemical control of feeding behaviour in the Dover sole *Solea solea*, *Comp. Biochem. Physiol.*, 73A, 89, 1982.

32. **Mackie, A. M.,** Identification of the gustatory feeding stimulants, in *Chemoreception in Fishes*, Hara, T. J., Ed., Elsevier, Amsterdam, 1982, 275.

33. **Person-Le Ruyet, J., Menu, B., Cadena-Roa, M., and Métailler, R.,** Use of expanded pellets supplemented with attractive chemical substances for the weaning of turbot (*Scophthalmus maximus*), *J. World Maricult. Soc.*, 14, 676, 1983.

34. **Cadena-Roa, M., Huelvan, C., Le Borgne, Y., and Métailler, R.,** Use of rehydratable extruded pellets and attractive substances for the weaning of sole (*Solea vulgaris*), *J. World Maricult. Soc.*, 13, 246, 1982.

35. **Métailler, R., Cadena-Roa, M., and Person-Le Ruyet, J.,** Chemical attractive substances for the weaning of Dover sole (*Solea vulgaris*): quantitative and qualitative approach, *J. World Maricult. Soc.*, 14, 679, 1983.

36. **Rubio-Rincon, E. A.,** Evolution de la Composition en Acides Gras de l'Ovocyte à la Larve de Turbot (*Psetta maxima* L.) en Fonction du Régime Alimentaire des Reproducteurs et des Larves Ainsi que de la Température d'Incubation, Thèse Doctorat de 3ème Cycle Université de Bretagne Occidentale, Brest, France, 1986.

GILTHEAD SEABREAM, *SPARUS AURATA*

George Wm. Kissil

INTRODUCTION

The gilthead seabream is a highly prized food fish found in the Mediterranean and Black Seas and along the eastern Atlantic Ocean from Senegal to the British Isles. It is most commonly found in coastal waters and hypersaline lagoons, where it lives out its life.[1] Natural spawning occurs during early winter, January to February in the eastern Mediterranean, after the adult populations leave lagoonal areas for deeper waters.

The first attempts at laboratory rearing of eggs and larvae were made in France and Italy in the early 1970s, with other countries joining these efforts in later years. Survival of larva to a stockable fry size of 1 g remained low into the 1980s, with only 0.1 to 16% being reported from Italy.[2] Low survival was attributed to the difficulty in finding proper food organisms for the relatively small size larva, 2.9 to 3.1 mm long at the first feeding stage of 3 to 4 days post hatching,[3,4] and an incomplete knowledge of the environmental requirements of the developing larva.

Early interest in gilthead development was overshadowed by greater interest in European seabass, which seemed less problematic. The larger size of first-feeding larvae reduced dependence on a live food chain and allowed for earlier weaning to dry diet formulations. The result was an increased effort to develop the seabass for commercial cultivation, leaving seabream development to advance at a slower pace.

During the 1980s there was a renewed interest in gilthead, as it was felt that seabass was well along in its development. Much information on environmental and larval nutritional requirements was produced,[5-12] which helped advance mass rearing of the seabream. As a result, a number of Mediterranean countries are presently involved in larval mass culture (Table 1). Production of market-size fish from this activity grew significantly from 1985 to 1987, from 110 to 590 tons,[13] reaching an estimated threefold increase over 1987 in 1989 based on juvenile production figures for 1988 (Table 1). It is speculated that production will reach at least 20,000 to 30,000 tons over the next decade due to the potential of the European market for this fish and the declining natural fishery of the Mediterranean.

Spawning and mass-rearing procedures of the gilthead seabream vary slightly throughout the Mediterranean region as a function of local R&D and differences in local conditions, but in general, these procedures are very similar. The following description represents a general procedure that is successfully used for mass rearing. Large quantities of fertilized eggs are obtained through the natural spawning of captive broodstock, usually with the help of hormones to ensure complete oogenesis and ovulation. Year-round spawning has been accomplished through the use of photoperiod and water-temperature manipulation[14] and is used in a number of countries — Israel, Italy, France, and Spain — to ensure a continuous supply of young fish for farmers. Broodstock, commonly used in a ratio of a number of males to each female, are allowed to spawn in tanks or pools and the floating fertilized eggs are collected from the outflowing water as it passes through a fine mesh net.

The fertilized eggs can develop into hatched larvae within 40 to 135 h in the range of water temperatures of 11 to 21.3°C.[15] At the preferred range for embryonic development, 15 to 20°C, development takes 45 to 51 h.[16,17] Hatched larvae are stocked into larval rearing tanks kept under controlled environmental conditions, where they are fed a regime of live food organisms supplemented with inert microdiets during the first month of life. The live food, rotifers, *Brachionus plicatilis*, and various growth stages of brine shrimp, *Artemia* sp., are

TABLE 1
Production of Juvenile Seabream for 1988
and Estimates for 1989 Production
in Thousands of Fish[a]

	1988	1989
Cyprus	(NA)[b]	(NA)
France	(NA)	(NA)
Greece	1200	1500
Israel	600	750
Italy	1100	2000
Portugal	(NA)	400
Spain	3000	5500
Tunisia	(NA)	(NA)
Turkey	500	1000

[a] From Report of the Working Group on Mass Rearing of Juvenile Marine Fish to the Mariculture Committee of ICES. Palavas-les-Flots, June 16–19, 1989. With permission.

[b] Quantity information not available (NA) but production underway.

raised on different unicellular algae, *Nannochloropsis* sp., *Isochrysis* sp., or *Tetraselmis* sp., and/or baker's yeast. Live food is usually enriched before being fed to the larvae with commercial products designed for this specific purpose or n-3 HUFA-rich emulsions and protein supplements prepared at each facility.

The first feeding with rotifers occurs 1 d before the onset of active feeding by the larva, 2 d after hatching. Newly hatched *Artemia* nauplii are added to the tanks 10 to 14 d after the start of rotifer feeding, and an overlap of the two is maintained to ensure that all larvae can obtain food of the proper size.

The use of microdiets during the first month of rearing is not widespread, although commercial products are available for use in combination with live food from as early as the first week after hatching. These early diets are designed as partial replacements for both rotifers and *Artemia*, having high protein, 55 to 60%, high lipid levels, 13 to 20+%, and being finely sized to ensure their consumption by small fish. Beyond the first month of rearing, fish (5 to 10 mg) are rapidly weaned from *Artemia* to a dry formulated larval or fry diet with continued supplementation of adult *Artemia*. Supplementation can continue up to 1 g weight, depending upon the availability of adult *Artemia*, the efficiency of the dry diet, and the desired survival.

Fish weighing 1 g are big enough for stocking, and the majority, 90%, are grown to commercial size (350 g) in floating cages located in protected areas. Growout can take over 18 months and will vary depending upon the water temperature in the different areas of culture. The formulated feeds provided during culture to commercial size will vary in protein and lipid level, protein usually decreasing as the fish grow. Protein levels of one commonly used commercial feed drop from a high of 56 to 46%, while lipids rise from 11 to 12% during growout.

LABORATORY CULTURE

Culture of gilthead on a laboratory scale would follow the same basic procedure outlined above. As a result of the large number of commercial hatcheries in operation today around the Mediterranean basin, it is likely that most research laboratories obtain their required life stages from such facilities.

TABLE 2
Examples of Semipurified Test Diets for
Gilthead Seabream

Ingredient (g/kg dry diet)	Diet 1[22]	Diet 2[22]
Casein	500	357.6
Fish protein conc.	—	21.0
L-arginine	—	10.8
L-cystine	—	5.8
L-methionine	—	5.9
L-phenylalanine	—	6.2
L-threonine	—	3.9
L-tryptophan	—	1.1
L-tyrosine	—	3.1
L-valine	—	2.7
Carboxymethyl cellulose	50	—
Corn starch	—	422.0
Dextrin	100	—
Glucose	50	—
Capelin fish oil	120	23.0
Soybean oil	—	57.0
Alphacel	110	—
Alginate	—	30.0
Attractant (Asp + Phe)	20	—
Mineral premix	10	10.0
Vitamin premix	40	40.0

A number of test diets have been successfully used to determine nutritional requirements in gilthead seabream.[18-21] Two of these used in studies to determine total protein, pyridoxine, and fatty acid requirements appear in Table 2, and practical diets used for energy studies, animal protein replacement trials, and the effects of attractants[23-25] are presented in Table 3.

NUTRIENT REQUIREMENTS

Although a minimal dietary protein level of 40% was established for optimal growth,[21] it is obvious from commercial seabream feed labels and personal observations that higher levels will provide better growth, especially in the earlier stages of development. Requirement levels for only four essential amino acids — arginine, lysine, methionine + cystine, and tryptophan — have been determined (Table 4) using graded levels of amino acids in test diets and determining the minimal level for nonretarded growth.[20] It is assumed that gilthead requires the same additional essential amino acids as other fish, but no requirement levels have been determined.

Lipid levels in diets for stockable fish (1 g) vary in the literature from 8 to 12%, with a recommendation for the use of fish oil as the major lipid source. These recommendations are based on an early study of the lipid needs of red seabream,[26] a close relative of gilthead, and subsequent work with the gilthead.[19,23,27] Seabream have been shown to grow better on fish oil in comparison to corn[19] or soybean oils,[27] which probably results from their inability to elongate linolenic acid to the n-3 HUFA, eicosopentaenoic (EPA), and docosahexaenoic (DHA) acids.[28]

Although no quantification of the n-3 HUFA needs for 1 g or larger seabream has been reported, larval requirements in the live food stages of rearing are being studied. Seabream fed n-3 HUFA- enriched rotifers and *Artemia* show better growth and survival and increased EPA

TABLE 3
Examples of Practical Diets for Gilthead Seabream

Ingredient (g/kg dry diet)	Diet 1[24]	Diet 2[25]
Fish meal	475.0	150
Meat and poultry meal	—	100
Soybean meal	—	450
Wheat flour	—	200
Wheat middlings	440.0	—
Fish oil	60.0	—
Soya oil	—	50
Mineral premix[a]	—	40
Vitamin premix	5.0[b]	10[c]
Pellet binder	20.0	—
Antioxidant	0.15	—

[a] To provide as g/kg diet: KI, 0.0016; $CaSO_4$, 0.0008; NaF, 0.04; $MnSO_4$, 0.12; $ZnSO_4$, 0.16; $CuSO_4$, 0.12; $FeSO_4$, 0.8; NaCl, 1.6; KCl, 3.6; CaCl, 8.6; $CaHPO_4$, 20; $MgCO_3$, 5.
[b] Source unknown.
[c] Commercial trout vitamin premix.

TABLE 4
Amino Acid Requirements of Gilthead Seabream

Amino acid	Requirement as % of dietary protein[20]
Lysine	5.0
Methionine + cystine	4.0
Tryptophan	0.6
Arginine	<2.6

and DHA content in their tissues. Levels of 8.5 mg n-3 HUFA/g dry weight of rotifers[29] and 30 mg/g dry weight of *Artemia* (Koven, unpublished data), although perhaps not optimum, have been shown to provide good growth and survival of seabream larva during the live food stage.

Vitamin premixes used for gilthead are based on salmonid formulations[21] and available information from red seabream.[30,31] The only study carried out with gilthead seabream indicated a pyridoxine requirement of 3 to 5 mg/kg dry diet to ensure unhindered growth and enzymatic activity associated with pyridoxine.[18]

No requirement values for carbohydrate, total energy, or minerals have been established for this fish. In addition, no digestibility information is available.

PRACTICAL DIETS

The two practical diet formulations appearing in Table 3 have been satisfactorily used to raise various sizes of seabream from 1 g. In spite of this fact, the lower protein levels of these diets (40 to 42%) would suggest that they are more appropriate for larger size seabream.

Practical diets are usually produced in a dry pelleted form, the size of which will vary with the fish's ability to consume the pellet. Fish are either fed by hand or by using various types of automatic or demand feeders. The type of feeding used will depend upon both economic and husbandry considerations.

TABLE 5
Example of a Fry Diet for Gilthead
Seabream

Ingredient	Percent
Powdered whole egg	61.0
Shrimp meal	6.0
Torula yeast	6.0
Lysine	2.3
Methionine	0.6
Vitamin premix[a]	0.3
Vitamin C	0.3
Choline chloride	0.3
Dicalcium phosphate	2.5
$CaCO_3$	1.9
Capelin oil	6.2
Dried squid	12.6

[a] Commercial trout vitamin premix.

SPECIAL DIETS

The literature does not provide examples of fry, juvenile, or broodstock diets for the seabream. Table 5 provides a diet that has proven successful in the rearing of juveniles on an experimental basis.

A common practice for broodstock populations is to provide natural foods, such as squid, shrimp, and various molluscs, in addition to their normal pelleted diets during the months of sexual maturation preceding spawning. Indications from the red seabream literature suggest positive effects of vitamin E, astaxanthin, and frozen raw krill supplements to the broodstock diet on egg viability and subsequent larval survival.[32] It is not known what factors are actually affecting egg quality, but it is suspected that antioxidative functions and essential fatty acid levels in the supplements may be bringing about the positive response.

REFERENCES

1. **Quignatd, J. P.,** Family Sparidae, in *FAO Species Identification Sheets for Fishery Purposes. Mediterranean and Black Sea,* Vol. 1, Fischer, W., Ed., FAO, Rome, 1971.
2. **Alessio, G., Gandolfi, G., and Schreiber, E. B.,** Techniche e metodiche generali di riproduzione artificiale dell'orata, *Sparus aurata* (L.) (Osteichthyes, Sparidae), *Inv. Pesq.,* 39, 417, 1975.
3. **Tandler, A.,** Overview: food for the larval stages of marine fish. Live or inert?, *Israel J. Zool.,* 33, 161, 1984/85.
4. **Kissil, G. Wm.,** Overview: rearing larval stages of marine fish on artificial diets, *Israel J. Zool.,* 33, 154, 1984/85.
5. **Divanach, P., Kentouri, M., and Dewavrin, G.,** The weaning and the development of biological performance of extensively reared sea bream, *Sparus aurata,* fry after replacing continuous feeders by self-feeding distributors, *Aquaculture,* 52, 21, 1986.
6. **Helps, S.,** An Examination of Prey Size Selection and its Subsequent Effect on Survival and Growth of Larval Gilthead Seabream (*Sparus aurata*), M.Sc. thesis, Plymouth Polytechnic, Plymouth, England, 1982.
7. **Pequin, C. L.,** The Effect of Photoperiod and Prey Density on the Growth and Survival of Larval Gilthead Seabream, *Sparus aurata* L. (Perciformes, Teleostei), M.Sc. thesis, Hebrew University, Jerusalem, Israel, 1984.
8. **Person-Le Ruyet, J. and Verillaud, P.,** Techniques d'elevage intensif de la daurade doree (*Sparus aurata* L.) de la naissance a l'age de deux mois, *Aquaculture,* 20, 351, 1980.

9. **Tandler, A. and Mason, C.,** Light and food density effects on growth and survival of larval gilthead seabream (*Sparus aurata*, Linnaeus, Sparidae) from hatching to metamorphosis in mass rearing systems, in *Proc. Warmwater Fish Cult. Workshop, World Maricult. Soc., Spec. Publ. Ser.*, 3, 103, 1983.

10. **Tandler, A. and Mason, C.,** The use of ^{14}C labelled rotifers (*Brachionus plicatilis*) in the larvae of gilthead seabream (*Sparus aurata*): measurements of the effect of rotifer concentration, the lighting regime and seabream larval age on their rate of rotifer ingestion, *Eur. Maricult. Soc.*, 8, 241, 1984.

11. **Tandler, A. and Helps, S.,** The effects of photoperiod and water exchange rate on growth and survival of gilthead seabream (*Sparus aurata*, Linnaeus: Sparidae) from hatching to metamorphosis in mass rearing systems, *Aquaculture*, 48, 71, 1985.

12. **Tandler, A., Har'el, M., Wilks, M., Levinson, A., Brickell, L., Christie, S., Avital, E., and Barr, Y.,** Effect of environmental temperature on survival, growth and population structure in the mass rearing of the gilthead seabream, *Sparus aurata*, *Aquaculture*, 78, 277, 1989.

13. **Popper, D. and Zohar, Y.,** Sea bream (*Sparus aurata*) and sea bass (*Dicentrarchus labrax*) culture in the Mediterranean region, in *Proc. Aquaculture Int. Cong. Expo.*, Vancouver, B. C., 1988, 319.

14. **Zohar, Y.,** Fish reproduction: its physiology and artificial manipulation, in *Fish Culture in Warm Water Systems: Problems and Trends*, Shilo, M. and Sarig, S., Eds., CRC Press, Boca Raton, FL, 1989, 259.

15. **Camus, P. and Koutsikopoulos, C.,** Incubation experimentale et developpement embryonnaire de la daurade royale, *Sparus aurata* (L.), a differentes temperatures, *Aquaculture*, 42, 177, 1984.

16. **Barnabe, G. and Rene, F.,** Reproduction controlee et production d'alevins chez la dorade *Sparus aurata* Linne 1758, *C. R. Acad. Sci., Paris*, 276, 1621, 1973.

17. **Alessio, G., Gandolfi, G., and Schreiber, E. B.,** Tecniche e metodiche generali di riproduzione artificiale dell'orata, *Sparus aurata* (L.) (Osteichthyes, Sparidae), *Inv. Pesq.*, 39, 417, 1975.

18. **Kissil, G. Wm., Cowey, C. B., Adron, J. W., and Richards, R. H.,** Pyridoxine requirements of the gilthead bream, *Sparus aurata*, *Aquaculture*, 23, 243, 1981.

19. **Koven, W. M. and Kissil, G. Wm.,** Requirement for ω3 polyunsaturated fatty acids in the gilthead seabream (*Sparus aurata*), *EMS Spec. Publ.*, No. 8, 93, 1984.

20. **Luquet, P. and Sabaut, J. J.,** Nutrition azotee et croissance chez la daurade et la truite, *Actes de Colloques, Colloques sur L'Aquaculture, Brest*, 1, 243, 1974.

21. **Sabaut, J. J. and Luquet, P.,** Nutritional requirements of the gilthead bream *Chrysophrys aurata*. Quantitative protein requirements, *Marine Biol.*, 18, 50, 1973.

22. **Kissil, G. Wm. and Koven, W. M.,** Comparison of test diets for the gilthead seabream (*Sparus aurata*), *Bamidgeh*, 39, 84, 1987.

23. **Marais, J. F. K. and Kissil, G. Wm.,** The influence of energy level on the feed intake, growth, food conversion and body composition of *Sparus aurata*, *Aquaculture*, 17, 203, 1979.

24. **Millan, L. M., Ortega, A., Alcazar, A. G., Lovell, R. T., and Posada, J. R.,** Optimum levels in substituting fish meal with soybean meal in practical diets for growing and fattening of gilthead seabream (*Sparus aurata*), *Animal Nutr. Highlights*, 1, 1989.

25. **Tandler, A., Berg, B. A., Kissil, G. Wm., and Mackie, A. M.,** Effect of food attractants on appetite and growth rate of gilthead bream, *Sparus aurata* L., *J. Fish Biol.*, 20, 673, 1982.

26. **Yone, Y., Furuichi, M., and Sakamoto, S.,** Studies on nutrition of red sea bream — III. Nutritive value and optimum content of lipids in diet, *Rep. Fish. Res. Lab., Kyushu Univ.*, 1, 49, 1971.

27. **Kissil, G. Wm., Meyers, S. P., Stickney, R. R., and Gropp, J.,** Protein/energy ratios in the feed of the gilthead bream (*Sparus aurata*), in *Proc. Warmwater Fish Cult. Workshop, WMS Spec. Publ.*, No. 3, 145, 1982.

28. **Kissil, G. Wm. and Lupatsch, I.,** Growth of gilthead seabream (*Sparus aurata*) and European seabass (*Dicentrarchus labrax*) on n-3 and n-6 rich lipids and their abilities to elongate and desaturate linoleic (18:2n-6) and linolenic (18:3n-3) acids, presented at Third Int. Symp. Feeding and Nutr. in Fish, Toba, Japan, August 28 to September 1, 1989.

29. **Koven, W. M., Tandler, A., Kissil, G. Wm., Freizlander, O., and Harel, M.,** The effect of level of dietary (n-3) highly unsaturated fatty acids on growth, survival and swim bladder development in *Sparus aurata* larvae, presented at Third Int. Symp. Feeding and Nutr. in Fish, Toba, Japan, August 28 to September 1, 1989.

30. **Yone, Y., Furuichi, M., and Shitanda, K.,** Vitamin requirements of red sea bream — I. Relationship between inositol requirements and glucose levels in the diet, *Bull. Jpn. Soc. Sci. Fish.*, 37, 149, 1971.

31. **Yone, Y. and Fujii, M.,** Studies on nutrition of red sea bream — X. Qualitative requirements for water-soluble vitamins, *Rep. Fish. Res. Lab., Kyushu Univ.*, 2, 25, 1974.

32. **Watanabe, T.,** Importance of the study of broodstock nutrition for further development of aquaculture, in *Nutrition and Feeding in Fish*, Cowey, C. B., Mackie, A. M., and Bell, J. G., Eds., Academic Press, London, 1985, 395.

GRASS CARP, *CTENOPHARYNGODON IDELLA*

Lin Ding

INTRODUCTION

Grass carp are naturally distributed in most rivers in China, except for the northwest region. They are also found from the Heilungkiang (Amur River) basin in eastern Asia to Thailand and Vietnam. But the area inhabited by grass carp has been greatly expanded in recent years. They have been introduced into the freshwaters of many regions in the world, such as central Asia, Japan, America, Europe, and the Arabian peninsula.[1,2] Grass carp is a very popular cultured fish, mainly because of its ability to feed on various plants. In addition, it grows rapidly to large sizes and has good meat quality. In pond culture, it can be polycultured with other fish.

Grass carp is one of the major freshwater species cultured in China. In recent years, grass carp culture techniques have been greatly improved, as illustrated in Figure 1. The area of fish ponds increased from about 16,600 ha in 1982 to about 17,300 ha in 1988. In these 7 years, the area increased by only 4.2%, whereas per unit yield of fish increased a great deal. In 1982, the per unit yield was 3975 kg/ha, of which 712.5 kg/ha was grass carp, or about 18% of the total yield. In 1988, the per unit yield increased to 7935 kg/ha, of which 2625 kg/ha was grass carp, or about 33% of total yield. The per unit yield has been doubled during this period of time. The yield of grass carp in some ponds has reached 7,500 kg/ha, suggesting a great yield in the future. One of the major reasons for this increase in production is the use of pelleted feed, which enables the higher density or net-cage monoculture of grass carp to be achieved.

Generally, grass carp pond culture can by divided into three growth stages. The first stage is fry rearing. At this stage, a monoculture is performed. The fry are reared from 3 to 4 d after hatching (about 7 mm in body length or 0.05 mg in weight) to a body length of 25 to 30 mm (about 250 mg). If natural food is sufficient, it takes 15 to 20 d to rear the fry. The second stage is fingerling rearing. In this stage, the fingerlings grow to 120 to 240 mm in length (about 14 to 100 g) in about 3 to 5 months. The third stage is adult fish culture, in which the fish are raised to marketable size of 1 to 2 kg or 4 to 5 kg in approximately 1 to 2 years.

LABORATORY CULTURE

Grass carp of different sizes can be used for experiments, depending on the purpose of the study. As a general rule, fingerlings are usually used for nutritional studies. Under laboratory culture conditions, the following precautions should be taken:

1. It is important to select healthy and uniform-sized juvenile fish from the pond. The acclimation of the fish in the laboratory is a long process. During the first 1 to 2 weeks, the fish lose weight, in spite of the fact that they are given feed. Then, there is a period of static weight. Finally, the fish gain weight again.
2. During the acclimation period, the fish should be fed a nutritionally complete diet. The diet may contain some drugs, such as sulfaguanidine or sulfathiazole, to prevent disease. Other drugs, such as diplerex or chlortetracycline, can be put in the water to treat the fish for external parasites.
3. After the acclimation period, the experiment can be started. The stocking density is usually 30 to 50 fish in each group, and the experiments are carried out in flow-through

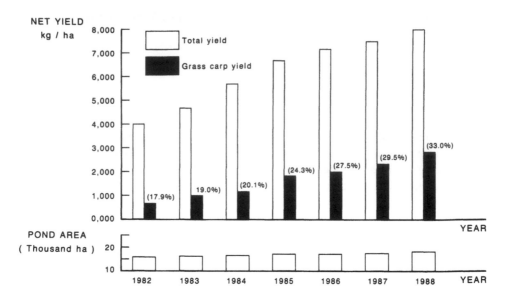

FIGURE 1. Annual net yield of freshwater pond cultured fish in Shunda County, Guangdong Province.

TABLE 1
Percentage Composition of Laboratory Diets

Ingredient	Fry diet[3]	Juvenile diet[4]
Casein	45	35
Dextrin		42
Hydrolyzed starch	45	
Alpha-starch		2
Fish oil	2	1
Soybean oil	3	1
Cellulose	1	5
Mineral mixture[a]	2	
Mineral mixture[b]		8
Vitamin mixture[18]	1	3
Vitamin C		0.5
Choline		0.5
Yeast	1	2

[a]   Salt mixture USP XIV no. 2 plus trace elements.
[b]   Salt mixture USP XII no. 2 plus trace elements.

aquaria. The fish are generally fed twice a day or several times over a 24-h period. Feeding rates are based on percent of body weight and are usually 2 to 4% or more, depending on the purpose of the experiment.

4.   During the feeding experiment, the water should be kept clean and fit for the growth of the fish. At least 70% of the water should be changed every day. All waste material should be removed. The oxygen level should be kept at 5 mg/l or higher. The optimal water temperature range is 22 to 28°C, and the pH should be between 7 and 8.

5.   Experimental diet formulas that have been used for nutritional research in laboratory culture of grass carp are presented in Table 1.

TABLE 2
Protein Requirements of Grass Carp

| Fish size (g) | Temperature (°C) | Protein requirement | | Ref. |
		% dry of feed	g/100 g body wt/day	
0.14–0.2	22–23	41–43		3
7–15	20.5–30	41.7	0.83	20
2.4–8.0	26–30.5	22.8–27.7	1.59–1.93	6
5.9–7.2	23–29	34.7–38.7	1.56–1.74	12
1.9	25–26	48.26	1.45–1.93	7
3.7	18–23	29.64	0.89–1.19	7
10.0	25	28.20	0.84–1.12	7

NUTRIENT REQUIREMENTS

PROTEIN AND AMINO ACIDS

Gross protein requirements have been determined for grass carp. Casein was used as the protein source, and levels of dietary protein were adjusted by adding different quantities of dextrin. Estimation of the optimum protein requirement was based on the relation between the protein content of the experimental diet and the total weight gain of the fish. The results showed a linear relationship between the percentage of protein in the diets and the increase in fish protein and weight up to optimal levels of 41 and 43%, respectively, average 42% for grass-carp fry with individuals weighing 0.14 to 0.2 g.[3] The same method was applied to grass-carp juveniles with body weights ranging from 2.4 to 8.0 g. The optimum dietary levels of protein ranged between 22.77 and 27.66%.[5,6] The optimum dietary protein requirements of grass-carp juveniles weighing 1.9, 3.7, and 10.0 g were found to be 48.26, 29.64, and 28.20%, respectively.[7]

The protein requirement of grass-carp juveniles was further studied by using the table of orthogonal arrays to ascertain the interaction between protein and other nutrients, such as carbohydrate, fat, minerals, and fiber. Results indicated that the daily protein requirement in the diet per 100 g of grass-carp juvenile weighing 5.87 to 7.15 g was 1.56 to 1.74 g, corresponding to an optimum dietary protein content of 34.66 to 38.66% when the feeding rate was 4.5%. Protein requirement values for grass carp are summarized in Table 2.

Table 2 shows that the optimum dietary protein requirement values range from 22 to 48%. These differences are due to differences in fish size, water temperature, and feeding rates. Other factors such as the experimental methodology and the interpretation of the results of the experiment may also affect these values.

Grass carp require the same 10 essential amino acids as other animals. Studies on quantitative amino acid requirements of fish were developed by Halver.[8] In recent years, however, many investigators have considered that the greatest proportion of body protein is in the form of muscle in young, growing fish. Therefore, it is reasonable to infer that the dietary essential amino acid requirement will be closely related to the amino acid profile of muscle protein.[9,10] Thus, the essential amino acid composition of grass carp muscle protein was analyzed and the results used as reference for the essential amino acid requirements of the fish (Table 3).

In Table 3 the ratio of essential amino acids in grass carp muscle is compared with the essential amino acid requirements of three other species of warmwater fish. Because the overall amino acid composition of muscle protein does not differ greatly among species, it is expected that there may be a good correlation between requirement and muscle patterns. The

TABLE 3
Relative Proportion of Essential Amino Acids in Grass Carp
Muscle Compared to Similar Proportions for Requirement
Values of Three Species of Fish[a]

Amino acid	Grass carp	Japanese[21] eel	Common[21] carp	Channel[21] catfish
Arginine	11.10	11.50	11.70	13.80
Histidine	4.71	5.40	5.80	4.90
Lysine	15.45	13.50	16.10	20.00
Phenylalanine	7.78	14.49	18.30	16.00
Leucine	16.52	13.50	9.50	11.20
Isoleucine	8.80	10.10	6.60	8.30
Methionine	4.49	8.10	8.80	7.50
Valine	11.50	10.10	10.20	9.50
Threonine	8.50	10.10	10.90	7.10
Tryptophan	2.34	2.70	2.20	0.90

[a] Expressed as weight of each essential amino acid as a percentage of the total weight of all essential amino acids.

agreement between ratios for most essential amino acids (arginine, histidine, lysine, isoleucine, valine, threonine, and tryptophan) is good, but it is variable for the other amino acids. Therefore, it seems reasonable to suggest that this type of information may be useful in designing test diets for fish when their amino acid requirements have not been established.

The data for the essential amino acid pattern of grass carp muscle have been computerized and used as a theoretical basis for formulating a balanced diet for grass carp. So far the method has proven feasible.[11]

LIPIDS

Dietary lipids serve both as sources of essential fatty acids (EFA) and energy. They also act as carriers for fat-soluble vitamins. The importance of EFAs for grass carp has not been studied and no requirement has been established. However, studies on the quantitative lipid requirement have been conducted and the results show the daily fat requirement to be 0.4 g/100 g body weight, corresponding to an optimum dietary fat content of 8% when the feeding rate was 5%.[12] Whereas it has been pointed out that grass carp do not utilize much lipid, thus an optimum dietary lipid level of 3.6% has been recommended.[13] The source of energy for grass carp is mainly carbohydrates or fat in the diet, but this needs to be further confirmed.

CARBOHYDRATE

Carbohydrates are the cheapest source of energy in fish diets. The results of an experiment at Zhongshan University indicated that grass carp can utilize a high content of dietary starch. The daily requirement of juvenile grass carp for carbohydrate (dextrin) was 1.12 g/100 g body weight, corresponding to a dietary carbohydrate level of 37 to 56% when the feeding rate was 2 to 3%.[12] Generally, fish cannot digest and utilize cellulose, however, grass carp can utilize a small amount of crude fiber. The digestion rate of labeled ^{14}C crude fiber fed to grass carp was 3 to 6%.[14] The crude fiber content in the diet for grass carp should be less than 15%.[5,12]

ENERGY

Providing an optimum energy level in the diet for fish is important, because an excess or deficiency of energy can result in reduced growth rates. However, data on energy requirements of grass carp are limited. The energy budget for grass carp has been reported. Fischer[15] showed that grass carp were hardly able to maintain their body weight when they were fed only on lettuce (*Lactuca sativa*). For fish ranging from 40 to 120 g, an average balanced equation at 22°C worked out to be 100 I = 16 M = 3 G + 81 E (where I = energy ingested; M = metabolism; G = growth; and E = excretion). When switched to a diet of Tubificidae, considerable improvement if growth occurred, resulting in an average equation of 100 I = 23 M + 17 G + 60 E. An experiment was conducted at Zhongshan University using whole chicken-egg powder in a test diet that contained 30% crude protein and 4500 kcal/kg. The experiment was carried out in flow-through aquaria at a controlled temperature of 28 ± 0.5°C. The fish were divided into two groups with average body weights of 24.70 ± 4.01 and 26.67 ± 6.25 g, respectively. The results showed that these grass carp obtained a higher growth energy, resulting in an average equation of 100 I = 36.91 M + 32.61 G + 30.48 E. It appears that it is necessary to supply a digestible energy (DE) of 3200 kcal/kg of test diet for grass-carp juvenile for maximum growth. Tentatively, the optimum dietary DE/protein ratio is recommended to be 10 kcal/g of protein for grass carp juveniles. The optimum DE/protein ratio for small common carp has been reported to be 8.3.[16]

VITAMINS

Only limited studies have been done with vitamin requirements and their respective deficiency signs in grass carp. The lack of vitamin C in the diet of grass carp causes hemorrhage in the eye and at the base of the pectoral and ventral fins. It was found that the vitamin C content in the liver increased with an increase of the vitamin content in the diet. The vitamin C requirement of grass carp is about 600 mg/kg diet when the vitamin C content in the liver and in the diet become constant.[17]

The requirements of other vitamins for grass carp have not been established. At present, the vitamin premixes used for nutritional research of grass carp in laboratory diets are usually referred to as Ogino's[18] or Halver's[8] formula.

MINERALS

Grass carp require the same minerals as warmblooded animals and fish for tissue formation and various metabolic processes. In addition, they utilize inorganic ions to maintain osmotic balance between fluids in their body and the environmental water. Minerals in the water can make significant contributions to the fish's requirement for some minerals, such as calcium. The total mineral requirement in the diet of grass carp was found to be 8 to 12%.[12] The orthogonal design method was applied to the mineral requirements of juvenile grass carp,[19] ranging in weight from 4.4 to 76.2 g, and the results showed that an adequate amount of the mineral mixture was 9.7% of dry diet. The mineral elements required in a diet to produce maximal growth are listed in Table 4.

The results also showed that the dietary calcium, phosphorus, sulfur, iron, magnesium, cobalt, and copper greatly affected the growth of grass carp fingerlings. The proper ratio between these main dietary elements was found to be approximately Ca:P:S:Fe:Mg = 18:12:9:2:1.

The composition of a typical mineral mixture for juvenile grass carp is shown in Table 5.

PRACTICAL DIET FORMULATION

The above-summarized requirement data provide a theoretical basis for practical diet formulation. Using these data and information on the costs and nutrient content of various feed ingredients, a least-cost commercial diet for grass carp can be formulated by computer.[11] Sample formulations are presented in Tables 6 and 7.

In China, most grass carp are cultured in ponds. In recent years, pelleted feeds have been widely used. Floating feeds are better than pelleted or extruded (nonfloating) feeds, but the

TABLE 4
Daily Mineral Requirements of Juvenile Grass Carp[a]

Mineral	Symbol	mg/100 g body wt/d
Calcium	Ca	32.6–36.7
Phosphorus	P	22.1–24.8
Potassium	K	25.0–28.3
Chlorine	Cl	20.9–23.5
Sulfur	S	15.5–17.4
Sodium	Na	7.7–8.7
Iron	Fe	4.1–4.6
Magnesium	Mg	1.8–2.0
Zinc	Zn	0.44–0.50
Cobalt	Co	0.04–0.05
Manganese	Mn	0.04–0.05
Copper	Cu	0.02–0.03
Iodine	I	0.005–0.006

[a] When the water contains calcium at levels of 20.2 to 28.6 mg/l and phosphorus at 0.005 to 0.023 mg/l.

From Huang, Y. and Liu, Y., *Acta Hydrobiol. Sinica,* 13, 134, 1989. With permission.

TABLE 5
Composition of a Typical Mineral Mixture[19]

Ingredient	Percent
Calcium biphosphate	12.287
Calcium lactate	47.424
Sodium biphosphate	4.203
Sodium chloride	3.233
Potassium sulfate	16.383
Potassium chloride	6.575
Ferrous sulfate	1.078
Ferric citrate	3.826
Magnesium sulfate	4.419
Zinc sulfate	0.474
Manganese sulfate	0.033
Cupric sulfate	0.022
Cobalt chloride	0.043
Potassium iodide	0.002

From Huang, Y. and Liu, Y., *Acta Hydrobiol. Sinica,* 13, 134, 1989. With permission.

cost of floating feeds are more expensive. Feeding procedures are affected by environmental factors, such as temperature and water quality; physical factors, such as rate of water exchange and type of rearing facility; management factors, such as frequency and rate of feeding; and type and size of fish. Grass carp have an optimum growth temperature around 30°C but grow at temperatures between 21 and 31°C. Below this temperature, feeding is erratic and daily feeding is usually uneconomical. The feeding rate of grass carp in ponds is recommended to

TABLE 6
Composition of a Typical 30% Crude Protein Fingerling Grass Carp Feed[a]

Ingredient	Percent
Fish meal (Menhaden or Peruvian)	4.0
Soybean meal	50.0
Peanut meal	4.0
Corn meal	15.8
Wheat meal or wheat feed	20.0
Vitamin mix[18]	0.2
Mineral mix[b]	4.0
Pellet binder	2.0

[a] All ingredients should be reground prior to pelleting. Pellet at as high a temperature as possible for maximum pellet hardness. Crumble pellets and sift out fines. For feeding in ponds from initial stocking of fingerlings through 30 to 100 mm in length.

[b] See Table 5.

TABLE 7
Grass Carp Feed Formulation Containing 24 to 26% Crude Protein Suitable for Pelleting or Extruding

Ingredient	Percent	
	A	B
Menhaden fish meal	5.0	2.0
Soybean meal	24.0	52.0
Corn meal	15.0	10.0
Wheat feed	55.0	10.0
Wheat middlings	—	22.0
Vitamin mix[18]	0.2	0.2
Mineral mix[a]	3.8	3.8
Pellet binder	2.0	—

[a] See Table 5.

be from 2 to 4% of the weight of the fish. Grass carp should be fed twice daily. The feeding should be started as soon as the water temperature warms up in the early spring until the water temperature begins to decrease in the late fall. A feeding rate of 1% of body weight is used on days when the water temperature is below 18°C in the winter.

Mechanical feeders include the demand type and the automatic type. These feeders have been used to increase yields in some larger fish farms in recent years in China.

REFERENCES

1. **Fischer, Z. and Lyakhnovich, V. P.,** Biology and bioenergetics of grass carp (*Ctenopharyngodon idella* Val.), *Pol. Arch. Hydrobiol.*, 20, 521, 1973.
2. **Stanley, J. G.,** Nitrogen and phosphorus balance of grass carp, *Ctenopharyngodon idella*, fed elodea, *Egeria densa*, *Trans. Am. Fish. Soc.*, 3, 587, 1974.
3. **Dabrowski, K.,** Protein requirement of grass carp fry (*Ctenopharyngodon idella* Val.), *Aquaculture*, 12, 63, 1977.
4. **Lin, D.,** Requirement of grass carp for essential amino acid, unpublished data, 1990.
5. **Liao, X., Lin, D., Mao, Y., and Cai, F.,** Experiments on enzymatic fiber pellets used as fish feed and protein requirements of grass carp (*Ctenopharyngodon idella* C. et V.), *J. Fish. China*, 4, 217, 1980.
6. **Lin, D., Mao, Y., and Cai, F.,** Experiments on the protein requirements of grass carp (*Ctenopharyngodon idella* C. et V.) juveniles, *Acta Hydrobiol. Sinica*, 7, 207, 1980.
7. **Liao, C. and Wang, Z.,** Studies on the requirement for dietary protein by grass carp fingerlings (*Ctenopharyngodon idella* C. et V.) in varied growth periods, *Freshwater Fish.*, 1, 1, 1987.
8. **Halver, J. E.,** Nutrition of salmonid fishes. III. Water soluble vitamin requirements of chinook salmon, *J. Nutr.*, 62, 225, 1957.
9. **Cowey, C. B. and Tacon, A. G. J.,** Fish nutrition-relevance to marine invertebrates, in *Proc. Second Int. Conf. Aquaculture Nutr.: Biochemical and Physiological Approaches to Shellfish Nutrition*, Pruder, G. D., Langdon, C. J., and Conklin, D. E., Eds., Louisiana State University, Baton Rouge, 1983, 13.
10. **Cho. C. Y., Cowey, C. B., and Watanabe, T.,** *Finfish Nutrition in Asia. Methodological Approaches to Research and Development*, International Development Research Centre, Ottawa, Publ. No. IDRC-233e, 1985, 154.
11. **Lin, D. and Mao, Y.,** Studies on computer formulation of fish feed, in *Chinese Symp. on Feedstuff Industry*, Chinese Assoc. Feedstuff Industry, 1986, 138.
12. **Mao, Y., Cai, F., and Lin, D.,** Studies on the daily requirements of protein, carbohydrate, fat, minerals and fiber of juvenile grass carp (*Ctenopharyngodon idella* C. et V.), *Trans. Chinese Ichthyol. Soc.*, 4, 81, 1985.
13. **Yong, W., Wang, Z., and Liao, C.,** Effect of fat content in the diet on the growth of grass carp, *Freshwater Fish.*, 6, 11, 1985.
14. **Zoology Division and Isotopes Laboratory,** Studies on the nutritional physiology of grass carp. III. On the application of ^{14}C to study the digestion and absorption of grass carp to crude cellulose, *Acta Scientiarum Naturlium Universitatis Sunyatsine*, 4, 106, 1978.
15. **Fischer, Z.,** The elements of energy balance in grass carp (*Ctenopharyngodon idella* Val.). Part II., *Pol. Arch. Hydrobiol.*, 17, 412, 1972.
16. **Takeuchi, T., Watanabe, T., and Ogino, C.,** Optimum ratio of dietary energy to protein for carp, *Bull. Jpn. Soc. Sci. Fish.*, 45, 983, 1979.
17. **Hu, Z., Wang, Z., and Liao, C.,** Vitamin C requirement of grass carp (*Ctenopharyngodon idella* C. et V.) in early growth, *Freshwater Fish.*, 2, 12, 1988.
18. **Ogino, C. and Saito, K.,** Protein nutrition in fish. I. The utilization of dietary protein by young carp, *Bull. Jpn. Soc. Sci. Fish.*, 26, 250, 1970.
19. **Huang, Y. and Liu, Y.,** Studies on the mineral requirement in juvenile grass carp (*Ctenopharyngodon idella* C. et V.), *Acta Hydrobiol. Sinica*, 13, 134, 1989.
20. **Chen, M-S. and Liu, H.-N.,** The utilization of dietary protein by young grass carp, *J. Fish. Soc. Taiwan*, 4, 67, 1976.
21. **National Research Council,** *Nutrient Requirements of Warmwater Fishes and Shellfish*, National Academy Press, Washington, D.C., 1983.

MILKFISH, *CHANOS CHANOS*

Chhorn Lim

INTRODUCTION

Milkfish, the only species known in the family of Chanidae, are widely distributed throughout the tropical and subtropical regions of the Indian and the Pacific Oceans.[1] This species is one of the finfish best suited for culture in the tropics because of its fast growth, efficient use of natural foods, herbivorous food habit, propensity to consume a variety of supplemental feeds, resistance to diseases and handling, and tolerance to a wide range of environmental conditions. Milkfish are euryhaline and can thrive in waters of 0 to 150‰ salinity.[2]

Milkfish are cultured on a large scale only in the Philippines, Indonesia, and Taiwan. Small-scale or experimental production is being practiced in some other Asian countries, such as Thailand, Malaysia, Vietnam, and Sri Lanka, and in Hawaii. Milkfish farming is believed to have begun in Indonesia some 700 years ago[3] and was introduced to the Philippines and Taiwan in the 16th century.[4] In 1983, milkfish was the single most important species produced through aquaculture in these countries, using more than 500,000 ha of brackish water and freshwater areas to produce over 365,000 t.[5] Throughout these countries, the general practice of milkfish farming in shallow brackish water ponds is similar but the yield varies considerably. The average annual production in Taiwan is about 2000 kg/ha[6] as compared to only 870 kg/ha in the Philippines[7] and 450 kg/ha in Indonesia.[8] The discrepancy between the production in these countries is mainly attributed to differences in skill and management inputs, such as stocking rate, size, pest and predator control, fertilization, and water management. Aside from its culture in brackish water ponds, milkfish have also been cultured extensively in freshwater pens in Laguna de Bay in the Philippines with an annual yield of about 4 t/ha.[9]

In recent years, milkfish production has been intensified through the use of more advanced technologies, such as multiple-size stocking, the modular pond system, and the deep-water method. The multiple-size stocking method presently used in the Philippines yielded an annual production of 2.2 to 2.7 t/ha.[10] The deep-water culture practiced in Taiwan with the use of formulated feeds produced 8 to 10 t/ha/year.[6]

LABORATORY CULTURE

Milkfish at various life stages have been reared successfully under laboratory conditions, using tanks or containers of different size and shape. Fry and fingerlings are the most common stages used in feeding studies. Fingerlings are generally reared in tanks equipped with flow-through systems and aeration. Non-flow-through systems provided with continuous aeration are used for fry. However, uneaten feed, feces, and debris should be removed daily, and approximately one half of the water should be changed at least once per day. Moreover, water current in the culture tanks must be minimized because milkfish tend to swim against the current.

Although milkfish are a very hardy fish, they must be acclimated to laboratory conditions before an experiment is begun. The duration of acclimation varies with fish size. The water and mineral balances of milkfish acclimated to different salinities generally stabilizes after 60 h for fingerling and only 24 h for fry.[11]

In the Philippines, decreasing salinity during holding is believed to reduce stress and fry mortality. However, at temperatures of 26 to 28°C, lowering salinity has been reported to have no effect on the survival of fry held for 2 weeks in plastic basins. On the other hand, at high

TABLE 1
Example of a Semipurified Test Diet
for Milkfish

Ingredient	Percent in diet
Casein, vitamin free	36
Gelatin	12
Dextrin	30
Marine fish oil	4
Corn or soybean oil	4
Celufil	5
Carboxymethyl celullose	2
Vitamin mix[a]	1
Mineral mix[b]	6

[a] Complete vitamin mix for warmwater fishes.[13]
[b] Mineral mix for purified diets.[13]

temperatures ($\geq$30°C), lowering salinity may be beneficial, since it has been observed that high temperature and high salinity ($\geq$ 30‰) induce early fry mortality.[12]

Milkfish at various life stages accept artificial feeds readily. Wild-caught fry (2 to 6 mg) and 14-day-old artificially bred fry (2.3 mg) have been reared successfully on artificial diets. A model formula of a semipurified diet is given in Table 1.

NUTRIENT REQUIREMENTS

PROTEIN AND AMINO ACIDS

The optimum dietary protein requirement for maximum growth, good feed efficiency, and survival has been reported to be about 40%.[14] No studies have been conducted on the amino acid requirements of milkfish. However, milkfish can be assumed to require the same 10 essential amino acids as other species. Leucine, lysine, and arginine may be the first limiting amino acids, since they occurred at high concentrations in the amino acid pattern of protein from the whole body of milkfish juveniles.[15] Milkfish appear to have the ability to utilize crystalline amino acids. Supplementation of 2.8% lysine hydrochloride to a corn-gluten-meal-based diet significantly improved the growth and feed efficiency of milkfish fry.[16]

Milkfish use proteins of animal origin better than plant proteins. Among animal proteins, fish meal and meat and bone meal have higher nutritive value than shrimp-head meal. Among plant proteins, soybean meal was superior to copra and *Leucena leucocephala* meals.[17] Vitamin-free casein supplemented with 0.5% L-tryptophan was a better protein source for milkfish fingerlings than casein and gelatin.[18]

The true protein digestibility of some feedstuffs by 60 and 175 g milkfish in fresh water and seawater is presented in Table 2. Regardless of fish size, gelatin had the highest digestibility value. Casein, fish meal, and defatted soybean meal were moderately digested and the digestibility coefficients tended to increase with fish size. *Leucena leucocephala* leaf meal was the least digestible. The digestibility of most feedstuffs tended to be lower in seawater than in fresh water.[20]

LIPIDS

Lipids are required in milkfish diets not only for energy but also as a source of essential fatty acids. Fish fed lipid-free or 7% lauric acid (LA) diets grew significantly poorer than those fed diets containing 6% LA plus 1% linoleic, or 0.5% linoleic and 0.5% linolenic acids.

TABLE 2

True Protein Digestibility of Some Feedstuffs by Two Sizes of Milkfish in Fresh Water (F) and Seawater (S)[19]

Feed ingredient	Water	Percent digestibility	
		60 g	175 g
Casein	F	84	88
	S	49	65
Gelatin	F	94	98
	S	98	97
Soybean meal	F	69	95
(defatted)	S	54	58
Fish meal	F	70	73
	S	52	71
L. leucocephala	F	40	42
Leaf meal	S	31	−10

However, the highest weight gain was obtained with fish fed the diet supplemented with 1% linolenic acid.[21] Thus, it appears that milkfish have a dietary requirement for linoleic and linolenic acids, but the optimum quantitative requirements are not known. Signs of essential fatty acid deficiency in milkfish were growth depression, increased levels of monoenic acids, decreased level of polyunsaturated fatty acids, and liver abnormalities, such as lipid infiltration in the blood vessels and cellular swelling.[21]

Because significant amounts of long-chain polyunsaturated fatty acids (PUFA) have been noted in milkfish livers, despite their absence in the natural food, milkfish may have the ability to bioconvert short-chain n-6 and n-3 fatty acids into long-chain n-6 and n-3 PUFA.[22] However, because milkfish grew better on a diet containing 20:4n-6 than on a diet containing 18:2n-6, there may be insufficient bioconversion of 18:2n-6 to 20:4n-6.[23]

Milkfish do not tolerate as high a level of dietary lipid as do salmonids. In studies using cod liver oil[24] and the combination of a 1:1 ratio of cod liver oil and corn oil,[25] a lipid level of 7 to 10% has been reported optimum for milkfish fingerlings. This level was sufficient to maintain liver structural and cellular integrity. Lipid levels below 7% resulted in decreased granulation and loss of nuclei of liver cells. Dietary lipid levels exceeding 10% caused minor disruption of hepatocytes from the formation of large lipid vacuoles, the loss of heptatic cord with development of fibrous tissues, and the occurrence of pyknotic nuclei.[24]

OTHER NUTRIENTS

No studies have been conducted on milkfish requirements for energy and carbohydrate. However, like other finfish, milkfish may not have a specific dietary carbohydrate requirement, but probably use carbohydrate as an energy source more efficiently than do coldwater fish. Diets containing up to 35% dextrin have been used successfully in experiments designed to determine other nutrient requirements. Commercial feeds in Taiwan used at present for pond feeding contain 45% or more of total carbohydrate.

No information is available on milkfish vitamin and mineral requirements. However, milkfish probably require the same vitamins and minerals as do other aquaculture species. Various vitamins and mineral mixes designed for coldwater and warmwater fishes have been used by different workers in milkfish nutrition research with satisfactory results. Thus, in the absence of information on these subjects, vitamin and mineral allowances established for other species are recommended.

FEEDS AND FEEDING

Milkfish is regarded as a herbivore. Its natural food and feeding habits have been extensively studied by a number of workers[26-36] and have been recently reviewed.[37-39] In natural habitats, milkfish feed on benthic, epiphytic, and planktonic organisms, including diatoms, copepods, gastropods, nematodes, lamellibranchs, and blue-green algae. Detritus also has been shown to constitute a large portion of the gut contents of milkfish. In extensive culture ponds, milkfish depend on two types of natural foods, locally known in the Philippines as lablab and lumut. Lablab, which appears as a greenish, brownish, or yellowish mat on the mud of the culture system, is a biological complex of small, bottom-dwelling plant and animal species consisting of diatom, blue-green algae, protozoans, nematodes, copepods, arthropods, rotifers, annelids, bivalves, and coelenterates. This group of foods is preferable to lumut, which consists mostly of filamentous green algae, such as *Chaetomorpha*, *Cladophora*, and *Enteromorpha*. Under laboratory conditions, during the first 2 to 3 weeks after hatching, the larvae are fed with a combination of such natural foods as rotifers, and *Artemia*.[40] However, 15-day-old fry have been successfully weaned abruptly to artificial diets.[41]

Milkfish find its food mainly by vision rather than chemosensory mechanisms. In the wild, milkfish fry feed mainly during daytime, with peak feeding activity at 0700 and 1900 h.[34] Under laboratory conditions, fry do not take brine shrimp in the dark. However, the ability to feed in the dark increases with growth, probably because of the development of chemosensory and auditory mechanisms. Juveniles were observed to feed on brine shrimp in the dark but did so less efficiently than in the light.[42] Larger fish (2 to 3 kg) take food day and night, but feeding activity was significantly less at night.[43]

Several studies have shown that milkfish from the fry to adult stages can be reared successfully with artificial diets. Before the development of compounded diets, during 1- to 2-week holding periods, milkfish fry were fed hard-boiled egg yolks, fine rice bran, or wheat flour. Such agricultural products or byproducts as rice bran, leaf meal, bread crumbs, or soybean meal are occasionally used for fish grown in nursery or growout ponds when there is insufficient growth or depletion of natural foods.

In recent years, efforts have been made to develop artificial diets for milkfish using available information on milkfish feeding habits and nutrient requirements, and information derived from other species. Feeds for fry have been formulated and used for fish grown in fresh water[44,45] and seawater.[46] An example of a practical diet formula for milkfish fry is given in Table 3.

Commercial feeds used at present for semiintensive pond culture in Taiwan contain 23 to 27% crude protein.[5] Since milkfish is an efficient feeder and feeds at the bottom of the food chain, natural pond foods make a valuable contribution to its nutrient requirements. Thus, these diets are assumed to be sufficient for satisfactory milkfish growth. A model formula of a practical pond feed for milkfish is presented in Table 4.

Very little is known about feeds and feeding of milkfish broodstock. Milkfish that matured and spawned naturally in floating cages at the SEAFDEC Aquaculture Department in the Philippines were fed with shrimp pellets containing 42% protein.[47] At Tung Hsing Hatchery, Pingtung, Taiwan, pond-reared broodstock were fed with a variety of feeds, including rice bran, wheat meal, soybean meal, and eel feed.[48] Purina Trout Chow or laboratory prepared diets containing 32 to 46% protein have been used successfully at the Oceanic Institute in Hawaii.[49] In Japara, Indonesia, a 36% protein feed in which 50% of the protein was supplied by fish meal has been used successfully for maturing broodstock reared in tanks.[50] Thus, in the absence of information on milkfish broodstock nutrient requirements, the diets for broodstock should contain a high level of good-quality protein. Also, the oil should be rich in polyunsaturated fatty acids. In addition, the diets should be fortified with extra vitamins and trace minerals. An example of a milkfish broodstock diet is given in Table 5.

TABLE 3
Example of a 40% Crude Protein Diet
for Milkfish Fry[46]

Ingredient	Percent in diet
Anchovy fish meal	30.0
Shrimp head meal	16.0
Soybean meal	20.0
Rice bran	11.5
Wheat flour	15.0
Cod liver oil	3.0
Vitamin mix[a]	1.0
Mineral mix[b]	3.5

[a] Complete vitamin mix for warmwater fishes.[13]
[b] One-half the amount of mineral mix for purified warmwater fish diets.[13]

TABLE 4
Model Formula of a Practical Pond (26% Protein)
Feed for Milkfish

Ingredient	Percent in diet
Fish meal, anchovy or menhaden	8.0
Soybean meal, 48% protein	31.5
Grains or grain byproduct	56.0
Pellet binder[a]	2.0
Dicalcium phosphate	1.5
Vitamin mix[b]	0.5
Mineral mix[c]	0.5

[a] Pellet binder may be hcmicellulose or lignin sulfonate.
[b] Vitamin mix for supplemental diets for warmwater fishes.[13]
[c] Mineral mix for practical diets for warmwater fishes.[13]

TABLE 5
Composition of a 36% Crude Protein Diet
for Milkfish Broodstock

Ingredient	Percent in diet
Fish meal, anchovy or menhaden	22.0
Soybean meal	35.0
Wheat flour	15.0
Corn meal	20.9
Fish oil	4.6
Dicalcium phosphate	1.0
Vitamin mix[a]	1.0
Trace mineral mix[b]	0.5

[a] Complete vitamin mix for warmwater fishes.[13]
[b] Mineral mix for practical diets for warmwater fishes.[13]

Milkfish accept a variety of feeds, in meal form and in moist, sinking or floating pellets. Crude feedstuffs are offered in meal form, whereas the compounded diets are usually processed into sinking pellets. Milkfish can use meal form feeds effectively. However, compounded feeds should be pelleted to minimize dissolution and separation of nutrients, and subsequent waste. Crude feed sources may be uneconomical when pelleted for pond feeding.

Pelleted feeds must have desirable physical characteristics, especially water stability and size. The feeds must remain water stable long enough to minimize nutrient loss and feed wastage. Hard and durable pellets are necessary when feeds are to be crumbled for feeding smaller fish. The most common pellet size used for feeding milkfish to marketable size (400 to 500 g) is approximately 4 mm in diameter and 6 to 8 mm long. Feeds in meal or crumbled forms of different particle sizes are used for fry and fingerlings.

Feeding rates for milkfish are affected by size, water quality (such as temperature, salinity, and dissolved oxygen), feeding frequency, and nutrient density of the diets, especially energy content. As with other fish, feed consumption rate of milkfish is inversely related to fish size. For example, with a diet containing 40% protein and 3450 kcal of ME/kg, a daily feeding rate of 20% of the biomass is optimum for 7.7 mg milkfish fry reared under laboratory conditions.[51] For fish averaging 0.60 g, feeding at 9% of the body weight resulted in a 130% increase in weight gain over the 5% feeding rate.[52] In pond environments where natural food is abundant, milkfish grown to marketable size are fed with commercial pellets containing 23 to 27% protein at a daily rate of 3 to 4% of body weight.[37]

Milkfish, like most other species, benefit from multiple daily feedings. The growth and feed efficiency of 0.6 g fingerlings fed at 5% or 9% of body weight increased by about 20% when the feeding frequency was increased from four to eight times daily.[52] Under pond conditions, milkfish are normally fed two to three times daily.[6] Feeds are offered to fish by hand or automatic feeders. The latter are commonly used in Taiwan. Automatic feeders equipped with two pipes extending toward the pond are installed on the dikes. The feeders contain devices that can be adjusted to deliver measured quantities of feed at given time intervals.

REFERENCES

1. **Chen, T. P.,** *Aquaculture Practice in Taiwan*, Fishing News Books Ltd., Surrey, England, 1976, chap. 1.
2. **Crear, D.,** Observations of the reproductive state of milkfish populations *Chanos chanos* from hypersaline ponds on Christmas Island (Pacific Ocean), in *Proc. World Maricul. Soc.*, Vol. 21, Avault, J. W., Jr., Ed., Louisiana State University, Baton Rouge, 1980, 548.
3. **Ronquillo, I. A.,** Biological studies on bangos (*Chanos chanos*), *Philip. J. Fish*, 9, 18, 1975.
4. **Ling, S. W.,** *Aquaculture in Southeast Asia — A Historical Overview*, University of Washington Press, Seattle, 1977, 108.
5. **Lee, C. S. and Banno, J.,** Milkfish culture and production in Southeast Asia — Present and future, presented at the Reg. Workshops on Milkfish Culture and Production in Southeast Asia — Present and Future, Tarawa, Kiribati, November 21 to 25, 1988, 24.
6. **Liao, I. C. and Chen, T. I.,** Milkfish culture methods in Southeast Asia, in *Aquaculture of Milkfish (Chanos chanos): State of the Art*, Lee, C. S., Gordon, M. S., and Watanabe, W. O., Eds., The Oceanic Institute, Waimanalo, Hawaii, 1986, 209.
7. **Sampson, E.,** The milkfish industry in the Philippines, in *Advances in Milkfish Biology and Culture*, Juario, J. V., Ferraris, R. P., and Benitez, L. V., Eds., Island Publishing House, Manila, 1984, 215.
8. **Chong, K. C., Poernomo, A., and Kasryno, F.,** Economic and technological aspects of the Indonesian milkfish industry, in *Advances in Milkfish Biology and Culture*, Juario, J. V., Ferraris, R. P., and Benitez, L. V., Eds., Island Publishing House, Manila, 1984, 199.

9. **Camacho, A. S. and Macalincag-Lagua, N.**, The Philippines aquaculture industry, in *Perspectives in Aquaculture Development in Southeast Asia and Japan*, Juario, J. V. and Benitez, L. V., Eds., Aquaculture Dept., SEAFDEC, Iloilo, Philippines, 1988, 91.

10. **Pamplona, S. D. and Mateo, R. T.**, Milkfish farming in the Philippines, in *Reproduction and Culture of Milkfish*, Lee, C. S. and Liao, I. C., Eds., The Oceanic Institute, Waimanalo, Hawaii, 1985, 141.

11. **Almendras, J. M. E.**, Changes in the Osmotic and Ionic Content of Milkfish Fry and Fingerlings During Transfer to Different Salinities, M.S. thesis, University of the Philippines, Iloilo, 1982.

12. **Villaluz, A. C., Villaver, W. R., and Salde, R. J.**, Milkfish Fry and Fingerling Industry of the Philippines: Methods and Practices, SEAFDEC Aquaculture Department Technical Report No. 9, Iloilo, Philippines, 1983, 84.

13. **National Research Council**, *Nutrient Requirements of Warmwater Fishes*, National Academy of Sciences, Washington, D.C., 1977, 78.

14. **Lim, C., Sukhawongs, S., and Pascual, F. P.**, A preliminary study on the protein requirements of *Chanos chanos* (Forsskal) fry in a controlled environment, *Aquaculture*, 2, 195, 1979.

15. **Coloso, R. M., Benitez, L. V., and Tiro, L. B.**, The effect of dietary protein-energy levels on growth and metabolism of milkfish (*Chanos chanos* Forsskal), *Comp. Biochem. Physiol.*, 89A, 11, 1988.

16. **Chiu, Y. N., Camacho, A. S., and Sastrillo, M. A. S.**, Effect of amino acid supplementation and vitamin level on the growth and survival of milkfish (*Chanos chanos*) fry, in *The First Asian Fisheries Forum*, Maclean, J. L., Dizon, L. B., and Hosillos, L. V., Eds., Asian Fisheries Society, Manila, 1986, 543.

17. **Samsi, S.**, Effects of Various Protein Sources on the Growth and Survival Rates of Milkfish (*Chanos chanos* Forsskal) Fingerlings in a Controlled Environment, M.S. thesis, University of the Philippines, Iloilo, 1979.

18. **Lee, D. L. and Liao, I. C.**, A preliminary study on the purified test diet for young milkfish *Chanos chanos*, in *Proc. International Milkfish Workshop Conference*, Iloilo, Philippines, May 19 to 22, 1976, 104.

19. **Ferraris, J. P., Catacutan, M. R., Mabelin, R. L., and Jazul, A. P.**, Effect of Fish Size and Salinity on Intestinal Passage Time and Protein Digestibility of Feedstuffs in Milkfish *Chanos chanos*, Technical Report, Aquaculture Department, SEAFDEC, Iloilo, Philippines, 1984.

20. **Ferraris, R. P., Catacutan, M. R., Mabelin, R. L., and Jazul, A. P.**, Digestibility in milkfish, *Chanos chanos* (Forsskal): effects of protein source, fish size and salinity, *Aquaculture*, 59, 93, 1986.

21. **Bautista, M. N. and De la Cruz, M. C.**, Linoleic (ω6) and linolenic (ω3) acids in the diet of fingerling milkfish (*Chanos chanos* Forsskal), *Aquaculture*, 71, 347, 1988.

22. **Benitez, L. V. and Gorriceta, I. R.**, Lipid composition of milkfish grown in ponds by traditional aquaculture, in *Finfish Nutrition in Asia — Methodological Approaches to Research and Development*, Cho, C. Y., Cowey, C. B., and Watanabe, T., Eds., IDRC-233e, Ottawa, Canada, 1983, 23.

23. **Kanazawa, A.**, Nutritional factors in fish reproduction, in *Reproduction and Culture of Milkfish*, Lee, C. S. and Liao, I. C., Eds., The Oceanic Institute, Waimanalo, Hawaii, 1985, 115.

24. **Alava, V. R. and De la Cruz, M. C.**, Quantitative dietary fat requirement of *Chanos chanos* fingerlings in a controlled environment, presented at the Sec. Int. Milkfish Aquaculture Conference, Iloilo, Philippines, October 4 to 8, 1983.

25. **Camacho, A. S. and Bien, N.**, Studies on the nutrient requirement of milkfish *Chanos chanos* (Forsskal), presented at The Tech. Symp. on Aquaculture, University of the Philippines, in The Visayas, February, 19, 1983, 17.

26. **Tampi, P. R. S.**, On the food of *Chanos chanos* (Forsskal), *Indian J. Fish.*, 5, 107, 1958.

27. **Schuster, W. H.**, Synopsis of Biological Data on Milkfish *Chanos chanos* (Forsskal), FAO Fisheries Biology Synopsis No. 4, FAO, Rome, Italy, 1960, 60.

28. **Rabanal, H. R.**, The culture of lablab, the natural food of milkfish or bangos, *Chanos chanos* (Forsskal) fry and fingerling under cultivation, *Phil. Fish J.*, 35, 22, 1966.

29. **Lin, S. Y.**, Milkfish Farming in Taiwan, A Review of Practice and Problems, Fish Culture Report No. 3, The Taiwan Fisheries Research Institute, Taiwan, 1968.

30. **Poernomo, A.**, Notes on food and feeding habits of milkfish (*Chanos chanos*) from the sea, in *Proc. International Milkfish Workshop Conf.*, Iloilo, Philippines, May 19 to 22, 1976, 162.

31. **Villaluz, A. C., Tiro, L. B., Ver, L. M., and Vanstone, W. E.**, Qualitative analysis of the contents of the anterior portion of the oesophagus from adult milkfish, *Chanos chanos*, captured in Pandan Bay, from May 10 to June 16, 1975, in *Proc. International Milkfish Workshop Conference*, Iloilo, Philippines, May 19 to 22, 1976, 228.

32. **Vicencio, Z. T.**, Studies on the feeding habits of milkfish *Chanos chanos* (Forsskal), *Fish. Res. J. Philipp.*, 2, 3, 1977.

33. **Guerrero, R. D., III**, Why lablab is nutritious for fish, *Mod. Agric. Ind. Asia*, 7, 12, 1979.

34. **Banno, J. E.**, The Food and Feeding Habit of the Milkfish Fry *Chanos chanos* (Forsskal) Collected from the Habitats Along the Coast of Hamtic, Atike, M.S. thesis, University of the Philippines in the Visayas, Iloilo, 1980, 77.

35. **Buri, P.**, Ecology on the feeding of milkfish fry and juveniles, *Chanos chanos* (Forsskal) in the Phillippines, *Mem. Fac. Fish., Kagoshima Univ.*, 1, 25, 1980.

36. **Kumagai, S. and Bagarinao, T. U.**, Studies on the habitat and food of juvenile milkfish in the wild, *Fish Res. J. Phil.*, 6, 1, 1981.

37. **Benitez, L. V.**, Milkfish nutrition, in *Milkfish Biology and Culture*, Juario, J. V., Ferraris, R. P., and Benitez, L. V., Eds., Island Publishing House, Manila, 1984, 133.

38. **Santiago, C. B.**, Nutrition and feeds, in *Aquaculture of Milkfish (Chanos chanos): State of the Art*, Lee, C. S., Gordon, M. S., and Watanabe, W. O., Eds., The Oceanic Institute, Waimanalo, Hawaii, 1986, 181.

39. **Watanabe, W. O.**, Larvae and larval culture, in *Aquaculture of Milkfish (Chanos chanos): State of the Art*, Lee, C. S., Gordon, M. S., and Watanabe, W. O., Eds., The Oceanic Institute, Waimanalo, Hawaii, 1986, 117.

40. **Juario, J. V., Duray, M. N., Duray, V. M., Nacario, J. F., and Almendras, J. M. E.**, Induced breeding and larval rearing experiments with milkfish *Chanos chanos* (Forsskal) in the Philippines, *Aquaculture*, 36, 61, 1984.

41. **Duray, M. and Bagarinao, T.**, Weaning of hatchery-bred milkfish larvae from live food to artificial diets, *Aquaculture*, 41, 325, 1984.

42. **Kawamura, G. and Hara, S.**, On the visual feeding of milkfish larvae and juveniles in captivity, *Bull. Jpn. Soc. Sci. Fish.*, 46, 1297, 1980.

43. **Kawamura, G. and Castillo, A., Jr.**, A new device for recording the feeding activities of milkfish, *Bull. Jpn. Soc. Sci. Fish.*, 47, 141, 1981.

44. **Santiago, C. B., Banes-Aldaba, M., and Sungalia, E. T.**, Effect of artificial diets on growth and survival of milkfish fry in fresh water, *Aquaculture*, 34, 247, 1983.

45. **Santiago, C. B., Pantastico, J. B., Baldia, S. F., and Reyes, O. S.**, Feeding for milkfish fingerling production in freshwater ponds, *Asian Aquaculture*, 11, 1, 1989.

46. **Alava, V. R. and Lim, C.**, Artificial diets for milkfish, *Chanos chanos* (Forsskal), fry reared in sea water, *Aquaculture*, 71, 339, 1988.

47. **Lacanilao, F. L. and Marte, C. L.**, Sexual maturation of milkfish in floating cages, *Asian Aquaculture*, 3, 4, 1980.

48. **Lin, L. T.**, My experience in artificial propagation of milkfish — Studies on natural spawning of pond reared broodstock, in *Reproduction and Culture of Milkfish*, Lee, C. S. and Liao, I. C., Eds., The Oceanic Institute, Waimanalo, Hawaii, 1985, 185.

49. **Kelley, C. and Lee, C. S.**, Artificial propagation, in *Aquaculture of Milkfish (Chanos chanos): State of the Art*, Lee, C. S., Gordon, M. S., and Watanabe, W. O., Eds., The Oceanic Institute, Waimanalo, Hawaii, 1986, 83.

50. **Poernomo, A., Lim, C., Vanstone, W. E., Daulay, T., and Anindiastuti**, Maturation of captive milkfish (*Chanos chanos*) in tanks, presented at the Workshop on Milkfish Reproduction, Tungkung, Taiwan, April 22 to 24, 1985.

51. **Lim, C.**, Effect of Feeding Rate on the Survival and Growth of Milkfish (*Chanos chanos*) Fry in a Controlled Environment. SEAFDEC Aquaculture Department Quarterly Research Report, II(4), 1978, 17.

52. **Chiu, Y. N., Sumagaysay, N. S., and Sastrillo, M. A. S.**, Effect of feeding frequency and feeding rate on the growth and feed efficiency of milkfish *Chanos chanos* Forsskal, juveniles, *Asian Fish. Sci.*, 1, 27, 1987.

PACIFIC SALMON, *ONCORHYNCHUS* spp.

Ronald W. Hardy

INTRODUCTION

In sheer numbers, Pacific salmon are among the most widely cultivated fish on Earth, with approximately 4 to 5 billion fry and juveniles reared each year by various state, federal, and private hatcheries and farms. Most of these fish are reared to the smolt stage, at which time they are physiologically and behaviorally prepared to migrate from fresh water to the sea. After release, the salmon grow to adulthood in the North Pacific Ocean. As the fish approach maturity, they undertake a second migration to nearshore areas, where most enter capture fisheries. Capture fisheries are managed to permit sufficient escapement of mature fish to rivers where they spawn to perpetuate the year-class.

There are five species of Pacific salmon native to North America and an additional species found only in Japan. The species differ somewhat in their life history, including the length of time spent in fresh water as juveniles. The North American species are chinook (*Oncorhynchus tshawytscha*), coho (*Oncorhynchus kisutch*), sockeye (*Oncorhynchus nerka*), pink (*Oncorhynchus keta*), and chum (*Oncorhynchus gorbuscha*). The cherry salmon (*Oncorhynchus masu*) is native only to Japan. Some of the differences in life history among the species as they pertain to hatchery rearing and capture fisheries are presented in Table 1. Recently, the scientific name of rainbow trout, *Salmo gairdneri*, was changed to *Oncorhynchus mykiss*, placing it in the same genus as the Pacific salmon. For the purposes of this chapter, rainbow trout are not included in the discussion.

The abundance of Pacific salmon in historical times is legendary, with estimates of the historical Indian salmon catch on the Columbia River of approximately 8,000,000 kg,[1] but the combined effects of overfishing and loss of freshwater habitat due to the construction of dams, destructive logging practices, and urbanization have greatly reduced the numbers of salmon in the Pacific Rim areas where salmon arc indigenous. In Alaska, careful management of escapement, coupled with enhancement of natural stocks by private, nonprofit hatcheries, has increased the salmon catch in recent years. Hatcheries in Alaska primarily rear and release pink and chum salmon, with annual releases totaling approximately 1 billion fish. In the more populated coastal states of the western United States, hatcheries are needed to supplement natural populations and, in many areas, fish of hatchery origin constitute approximately half of the commercial and sport catch. Approximately 750 million smolts, mainly chinook and coho salmon, are released from hatcheries in this region. In Japan, approximately 2 billion chum fry are released annually, yielding 50 to 55 million adults returning to the capture fisheries.[2] Russia releases an estimated 1 billion salmon into North Pacific waters each year.

Chinook and coho are the species of salmon that have been reared in hatcheries for the longest time; the first chinook hatchery was built in California in 1872. In the past, construction of hatcheries was justified by the observation that the hatching rates of hand-spawned eggs were generally over 90%, whereas the hatching rates of naturally spawned eggs were in the order of 5 to 25%, depending upon predation rates, flooding, and density of spawners on the spawning grounds. Early hatcheries were notable for their lack of success in rehabilitating overfished stocks of salmon, which is not surprising considering that most salmon were released as unfed fry. Investigations by early biologists suggested that releasing hatchery salmon at an approximate size and time that wild smolts normally migrate to the sea would be a beneficial practice, and this approach soon became standard. With extended rearing came the need to feed the fish until release and concurrently the need to identify the nutrient

TABLE 1
Salmon Life History Characteristics[a]

Pacific salmon[b]	Size at hatchery release or saltwater transfer (g)	Duration of freshwater residence (months)	Size at maturity (kg)	Age at maturity (years)
Chinook (king)	5–10	3–5	1–55	3–5
Coho (silver)	25–40	15–20	3–6	2–4
Chum (dog)	0.5–2	0.2	1–20	3–4
Pink	0.5–2	0.2	1.4	2
Sockeye (red)	5–25	12–24	2–4	2–4
Masu (cherry)[c]	10–22	15–20	3–5	3

[a] Exceptions to these characteristics exist in certain populations.
[b] Other common name in parenthesis.
[c] Only females migrate to the sea. Males remain in rivers and generally do not exceed 1 kg in weight.

requirements of feed. One of the first systematic studies of the nutritional requirements of salmonids was conducted by Embody and Gordon,[3] who caught wild trout, enumerated the stomach contents, and conducted chemical analyses of representative insects and other food items. They found the proximate composition of the diet of wild trout to be approximately 49% protein, 15 to 16% fat, 8% fiber, and 10% ash. The accuracy of their work is illustrated by the fact that most salmonid feeds manufactured today have a similar proximate composition.

From the turn of the century until the mid-1950s, feeds were formulated for hatchery salmon empirically, because insufficient information regarding the nutritional requirements of the fish was available.[4] Often the dietary ingredients used were selected solely because they were locally available. The development of a suitable purified diet for salmon by Halver[5] provided the missing element that had impeded research on salmon nutritional requirements, and nearly all of the information now used to formulate feeds has been developed since then. Concurrent with Halver's work was the development by the Oregon Fish Commission of the Oregon moist pellet (OMP), which was developed primarily to eliminate the transmission of fish tuberculosis from feed to fish resulting from the use of spawned salmon carcasses as one of the ingredients in hatchery diets.[6] The development of the OMP resulted in centralization of feed manufacturing by private companies, eliminating the need for each salmon hatchery to prepare its own feed. The success of the OMP provided the foundation for the subsequent expansion of the Pacific hatchery system and ultimately for the salmon aquaculture industry. Early efforts to raise salmon in marine net pens used the OMP as the standard diet, even though it was developed as a feed for fry and juvenile salmon. The formulation satisfied the nutritional requirements and supported the growth of postjuvenile and maturing salmon. Advances made in the formulation and manufacture of dry compressed and dry extruded diets for salmonids based on sound nutritional information eventually made the OMP uneconomical to use as a feed for postjuvenile salmon, although it is still widely used in freshwater hatcheries as a fry and juvenile diet, and formulations derived from it are used in Japan in seawater cage culture of coho salmon and other marine species.

LABORATORY CULTURE

Pacific salmon are relatively easy to rear in a laboratory setting as long as certain environmental conditions are met. These include fresh water of 8 to 15°C that is fully saturated with oxygen and sufficient water flow and rearing space to satisfy the density needs of the fish.

Rearing density in an experimental setting is generally lower than that in a production setting, with typical rearing densities of 2 to 4 kg/m³. In an experimental laboratory, care must also be taken to ensure that rearing densities are not too low, and that a sufficient number of fish are used per tank. Fry and juvenile Pacific salmon are territorial, and hierarchical systems will develop if too few fish are placed in each tank. Generally, 50 to 100 fish is the minimum number of fish to use per tank in nutritional studies to avoid the confounding effects of hierarchical behavior and also to increase "feelings of security" among the fish, which seem to appreciate being part of a small school. With postjuvenile fish, the numbers per tank can be reduced to 30 to 50.

Pacific salmon respond dramatically to changes in photoperiod, and care must be taken to control or account for photoperiod in the laboratory. Photoperiod can greatly influence a diet study by triggering smoltification in juveniles, as is the case with an increasing photoperiod; by reducing growth, as is the case with a decreasing photoperiod; or by accelerating or retarding maturation. As in all laboratory experiments, care should be taken to equalize environmental variables, such as light intensity, water flow, handling, disease treatment, and human traffic. If equalization is impossible, then an appropriate experimental design must be employed to minimize the effects of variation in performance due to extraneous variables on the performance response due to the dietary treatment.

Semipurified diet development for Pacific salmon had its genesis in the efforts of McLaren et al.[7] and Wolfe,[8] who were the first to test such diets for rainbow trout. The diets of these investigators were fairly successful, but their own experimental evidence showed that their diets were not truly vitamin free. Halver and Coates[9] developed the first truly vitamin-free diet for salmonids and this diet formulation, called H440 in the 1973 NRC bulletin,[10] has since been modified by other investigators to take into account new information on the nutrient requirements of salmonids.

Modification of H440 by eliminating casein and gelatin and replacing them with a mixture of crystalline amino acids resulted in a diet formulation suitable for determining the amino acid requirements of Pacific salmon.[11] The original H440 diet formulation, and several more recent modifications, are shown in Table 2. The first principal modification of the H440 diet took place when a portion of the corn oil was replaced with fish oil.[11] When information began to accumulate regarding the amino acid requirements of Pacific salmon, it was quickly noticed that casein/gelatin-based diets were deficient in arginine. Addition of an amino acid premix greatly increased the nutritional quality of purified diets.

Additional modification of H440 occurred when attempts were made to modify the amino acid composition of the diet to approximate whole chicken eggs, whole fish eggs, or whole fish.[12] Today, many modifications of H440 exist that are used in instances when specific nutrient requirements are being studied. Overall, H440 and its derivatives have been used as the test diet to study dietary protein, lipid and carbohydrate levels, vitamin requirements, amino acid requirements, fatty acid requirements, and carotenoid requirements for pigmentation. Casein/gelatin-based diets have also been used to investigate the mineral requirements of salmonids, but the relatively high content of phosphorus and zinc in casein/gelatin-based diets make them unsuitable for determining the dietary requirements of these minerals. Egg white, blood fibrin, or highly refined fish muscle protein appear to be acceptable protein replacements for casein in cases where the residual mineral levels of casein are too high.[13,14]

Growth of Pacific salmon fed casein/gelatin-based diets has been generally lower than that of fish fed practical diets, but refinements in the formulation to take into account new information on the nutrient requirements of Pacific salmon have reduced this difference. For example, Halver[5] reported percent body weight increases per day of 0.79% in fingerling chinook salmon reared in 8°C water. Percent body weight increases per day of chinook salmon reared in production hatcheries are approximately 2% in similar water temperatures. Later publications show that substituting herring oil for corn oil in H440 and reducing the protein

TABLE 2
Composition of Test Diets for Pacific Salmon

Ingredient	Halver's 1st diet	H440	Amino acid test diet	Modified diet
Casein, vitamin-free	52.88	38.0	—	40.8
Gelatin	14.69	12.0	—	8.0
Dextrin	7.83	28.0	6.0	16.0
Corn oil	8.8	6.0	5.0	—
Cod liver oil	—	3.0	2.0	—
Carboxymethyl cellulose	—	—	10.0	—
Alpha-cellulose	8.8	9.0	—	4.7
Herring oil	—	—	—	15.0
Mineral mixture	3.92[a]	4.0[c]	4.0[c]	8.0[e]
DL-methionine	0.98	—	—	—
L-tryptophan	0.49	—	—	—
Vitamin mixture	1.59[b]	1.0[d]	3.0[d]	2.0[f]
Amino acid mixture	—	—	70.0	4.4
Ascorbic acid	—	—	—	0.1
Choline chloride (70%)	—	—	—	1.0

[a] Supplied the following per kg dry diet: USP XII salt mixture no. 2, 38.94 g; aluminum chloride, 7.05 mg; zinc sulfate, 139 mg; cuprous chloride, 4.3 mg; manganous sulfate, 3.11 mg; potassium iodide, 0.67 mg; cobalt chloride, 4.09 mg.

[b] Supplied the following as mg per kg dry diet: thiamine hydrochloride, 60; riboflavin, 200; pyridoxine hydrochloride, 40; nicotinic acid, 800; calcium pantothenate, 28; inositol, 4000; biotin, 6; folic acid, 15; para-aminobenzoic acid, 400; choline chloride, 8000; ascorbic acid, 2000; alpha-tocopherol, 400; menadione, 40; beta-carotene, 12; activated 7-dehydrocholesterol, 0.045; and vitamin B_{12}, 0.09.

[c] Supplied the following per 1 g dry diet: calcium biphosphate, 5.43 g; calcium lactate, 13.976 g; ferric citrate, 1.188 g; magnesium sulfate, 5.28 g; potassium phosphate (dibasic), 9.592 g; sodium biphosphate, 3.488 g; sodium chloride, 1.74 g; $AlCl_3 \cdot 6H_2O$, 6 mg; KI, 6 mg; $CuCl_2$, 4 mg; $MnSO_4 \cdot H_2O$, 32 mg; $CoCl_2 \cdot H_2O$, 40 mg; and $ZnSO_4 \cdot H_2O$, 120 mg.

[d] Supplied the following as mg per kg dry diet: thiamine hydrochloride, 50; riboflavin, 200; pyridoxine hydrochloride, 50; choline chloride, 5000; nicotinic acid, 750; calcium pantothenate, 500; inositol, 2000; biotin, 5; folic acid, 15; ascorbic acid, 1000; vitamin B_{12}, 0.1; menadione, 40; and α-tocopherol acetate, 400.

[e] Supplied the following per kg dry diet: alpha-cellulose, 33.12 g; $CaHPO_4$, 29.28 g; MgO, 1.6 g; and $NaHPO_4$, 16 g. Trace mineral solution was added with water to supply the following per kg dry diet: $CoCl_3 \cdot 6H_2O$, 4 mg; $MnSO_4 \cdot H_2O$, 25 mg; $ZnSO_4 \cdot 7H_2O$, 163 mg; and $NaSO_4$, 2.4 mg.

[f] Supplied the following as mg per kg dry diet: thiamine mononitrate, 104; riboflavin, 222; pyridoxine hydrochloride, 41; niacinamide, 586, calcium D-pantothenate, 564; folic acid, 34, and biotin 1.58.

content increased percent body weight increases per day to 1.6% in fingerling chinook salmon.[15] Recent laboratory work has shown that chinook salmon fry reared in 10°C water can achieve percent body weight increases of slightly over 3% per day when fed semipurified diets. This is identical to the percent body weight increases obtained when feeding a practical diet in similar circumstances.[16] Growth rates of Pacific salmon fed purified diets in which a substantial proportion of dietary protein has been replaced with mixtures of crystalline amino acids are generally lower than those of fish fed semipurified diets containing intact protein, i.e., casein/gelatin, purified fish protein, or fish fed practical diets.

NUTRIENT REQUIREMENTS

Information on the nutrient requirements of Pacific salmon appears to be relatively complete compared to the information known for most other species of farmed fish. In fact, casual observers frequently have the impression that the nutrient requirements of Pacific salmon are completely known, or at least enough well known to permit least-cost feed formulation.

However, examination of the literature quickly reveals several significant points that challenge that assumption. First, nearly all of the information available on nutrient requirements of Pacific salmon was determined in the 1950s and 1960s, when many of the purified diets in use at the time were not completely balanced due to incomplete information on salmon nutrient ingredients.

Second, nearly all of the research conducted to determine the nutrient requirement of Pacific salmon has been conducted using fry or fingerlings. This is logical because, prior to the early 1970s, Pacific salmon rearing was limited to rearing fish to the smolt stage for release to enhance the fishery. Rearing of postjuvenile Pacific salmon to market size is a relatively recent practice.

Third, despite the fact that there are six species of Pacific salmon with quite different life histories, most information on nutrient requirements of Pacific salmon has only been determined using juvenile chinook salmon. These requirements have been applied to the other species of Pacific salmon, a practice that appears to be justified when comparing the qualitative nutrient requirements, but may not be justified when comparing the quantitative nutrient requirements.

Fourth, early work usually did not include the degree of statistical evaluation of the results that is now considered necessary, making it difficult to evaluate the results and decide which dietary treatments were significantly different from others at an acceptable level of probability. Finally, production of practical diets for postjuvenile Pacific salmon, at least in North America, has been greatly influenced by production and formulation technologies developed for Atlantic salmon in Europe and, to a lesser extent, with rainbow trout in North America. Anecdotal evidence suggests that feed formulation practices that are successful with these species may not always be appropriate for Pacific salmon.

PROTEIN AND AMINO ACIDS

Various investigators have determined the protein requirement for several species of Pacific salmon in experimental settings (Table 3). The apparent protein requirement is influenced by a number of factors, including the size of the fish, the dietary energy content, and the availability of individual amino acids from dietary ingredients.[17] Zeitoun et al.[18] found that rainbow trout reared in seawater had a higher protein requirement than equivalent fish reared in fresh water, and this finding has been assumed to be true for Pacific salmon as well, although published evidence to support it is lacking. Water temperature is also reported to influence the dietary protein requirement of juvenile chinook salmon. Delong et al.[19] observed that chinook salmon required 40% dietary protein when reared in 8°C water, and 55% dietary protein when reared in 15°C water, but this observation has not been confirmed by other investigators.

Protein source and quality is an important consideration in Pacific salmon feed formulation, particularly in fry and juvenile fish. It is generally accepted that the quality of fish meal, particularly the drying temperature used in fish meal manufacture, greatly affects its nutritional value to juvenile salmonids.[20] Pacific salmon diets are generally formulated to contain high levels of fish meal, and efforts to replace the fish meal with other protein sources, such as soybean meal, have been unsuccessful.[21]

Extraction of phenolic compounds, which are thought to reduce the palatability of soybean products in the diets of juvenile salmon, and supplementation of diets containing soybean meal with amino acids have improved growth rates, but not to acceptable levels.[22] Diets containing soybean meal have been used commercially to rear postjuvenile salmonids, indicating that aversion to soybean meal in the diet may be reduced in larger fish. Currently, research efforts are increasing in the areas of alternate protein sources for Pacific salmon and the effects of fish meal quality on fish growth.

Pacific salmon require the same 10 amino acids required by other fish,[16,23] but the

TABLE 3
Estimated Protein Requirements of Pacific Salmon

Growth stage	Percent of diet
Fry	45–50
Juveniles	40
Postjuveniles	40
Broodstock	45

TABLE 4
Essential Amino Acid Requirements of Pacific Salmon[a]

Amino acid	Chinook	Coho	Sockeye	Chum
Arginine	6.0[53]	5.8[53]		
Histidine	1.8[53]	1.8[53]		1.8[54]
Isoleucine	2.2[24]			
Leucine	3.9[24]			
Lysine	5.0[55]			4.8[54]
Methionine	4.0[57]			
Phenylalanine	5.1[24]			
Threonine	2.2[56]			3.0[54]
Tryptophan	0.5[58]	0.5[58]	0.5[58]	0.7[59]
Valine	3.2[24]			

Expressed as a percent of dietary protein. Superscript numbers indicate reference.

quantitative requirements for several amino acids are slightly different (Table 4). There is some evidence to support an interaction between isoleucine and leucine, which results in an increase in the dietary isoleucine requirement with increased dietary levels of leucine.[24] The dietary amino acid requirements have all been determined using fry and juvenile salmon, and these values are used to formulate diets for all sizes and ages of fish. One interesting observation regarding sockeye salmon is that dietary tryptophan deficiency resulted in scoliosis and lordosis that was reversible when tryptophan was restored to the diet.[25] This condition has also been observed in chum salmon fry fed diets deficient in tryptophan, and the condition was prevented when 5-hydroxy-L-tryptophan was added to the diet.[26] This condition has not been observed in chinook or coho salmon.

LIPIDS AND ESSENTIAL FATTY ACIDS

Coho salmon have been reported to require between 1 and 2% n-3 fatty acids in the diet to prevent deficiency signs and to sustain normal growth rates.[27] Takeuchi et al.[28] reported that chum salmon fry require 1% n-3 fatty acids in the diet when the n-3 fatty acid source was linolenic acid, but that the requirement was approximately 0.5% when longer chain n-3 fatty acids were supplied. The essentiality of n-6 fatty acids is not yet certain. Takeuchi et al.[2] reported that chum salmon fry grew best when fed a diet containing both 1% linoleic acid and 1% linolenic acid, while Yu and Sinnhuber[27] provided evidence that growth of coho salmon fed diets containing over 1% linoleic acid was reduced. Both of the studies mentioned above used methyl esters as the dietary lipid. Recent work with commercial oils suggest that the growth rates of Pacific salmon are not reduced by high levels of n-6 fatty acids. Dosanjh et al.[29] fed juvenile chinook salmon diets containing canola oil, pork lard, and herring oil, both singly and in combination for 62 d, and did not observe reduced growth in fish fed diets containing up to 16.2% n-6 fatty acids, nor was seawater survival and growth influenced by dietary oil source.

The fatty acids of Pacific salmon can be modified to some extent by feeding diets with varying fatty acid composition. Monoenes, dienes, and polyunsaturated fatty acid levels in the whole body and in the flesh of Pacific salmon respond to changes in dietary fatty acid concentration.[29-32] Saturated fatty acids respond somewhat to dietary fatty acid level, but generally remain within a narrow range, regardless of diet. As long as the essential fatty acid requirement is met in the diet, the overall fatty acid composition of the diet does not appear to influence growth rates or the health of Pacific salmon. Although the fatty acid composition of broodstock coho salmon and their gametes reflects the fatty acid composition of the diet, there is no evidence that the quality of the gametes is influenced by dietary fatty acid composition, providing that the essential fatty acid level in the diet is above 1%.[33]

CARBOHYDRATES

Carbohydrates are not essential nutrients for Pacific salmon, but they deserve mention as a source of dietary energy. The only published report concerning carbohydrates in Pacific salmon is that of Buhler and Halver,[34] who fed chinook salmon fry purified diets containing dextrin, fructose, galactose, glucosamine, glucose, maltose, potato starch, or sucrose at 20% of the diet. After 14 weeks of feeding, fish fed the diets containing glucose, maltose, or sucrose weighed more than those fed diets containing dextrin or fructose, which in turn weighed more than fish fed diets containing galactose, glucosamine, or potato starch. They concluded that the nutritive value of the carbohydrate sources to the chinook salmon was inversely proportional to the chemical complexity of the carbohydrate source. The growth rate of the group of fish showing the best growth in that study was 1.13% per day. In another part of the same study, chinook salmon fry were fed purified diets for 18 weeks containing levels of dextrin and α-cellulose from 0 to 48% of the diet. The results showed that replacing α-cellulose with dextrin in the diets increased the growth of the fish, with the diet containing 48% dextrin and 0% α-cellulose supporting the highest growth. Varying the dextrin level in diets without α-cellulose resulted in the conclusion that a dietary level of 20% dextrin was optimal for supporting growth. As mentioned earlier, the lack of statistical evaluation of the growth results in the literature from the 1950s and 1960s makes it difficult to evaluate the finer points presented in most publications, and complicates the interpretation of the results of various carbohydrate levels and sources on growth in Pacific salmon.

VITAMINS

The qualitative vitamin requirements are similar among Pacific salmon studied to date, and there is little evidence to suggest that further research will reveal any major differences among species.[5,35] Some differences may exist, however, in the quantitative vitamin requirements of Pacific salmon at different life stages, in different rearing environments, or with diets of varying protein/energy ratios. Previous nutritional history, especially among large fish, has a profound effect on the dietary needs of salmonids for essential fat-soluble vitamins and, surprisingly, for certain water-soluble vitamins. Some evidence exists to suggest that dietary vitamin requirements depend on growth rates and feed conversion ratios of the feed being used, with rapidly growing fry and fingerlings possibly requiring higher dietary levels than large fish that grow more slowly. Dietary stability of vitamins in the diets of Pacific salmon has not been thoroughly studied, but is likely not substantially different in dry salmon diets from other fish diets or pet food. Moist diets present a more challenging environment for stability of certain vitamins, including vitamin E, ascorbic acid, and possibly thiamine in diets containing unpasteurized fish waste. The use of protected forms of vitamin E and ascorbic acid, or overfortification to ensure adequate intake by the fish after processing and storage losses occur, are proven ways to overcome the problems with vitamin stability in moist feeds.

Most of the research on quantitative vitamin requirements of Pacific salmon has been conducted using chinook salmon fry fed purified diets containing graded levels of the various

TABLE 5
Vitamin Requirements of Pacific Salmon

Vitamin	Requirement (per kg dry diet)	Basis of requirement
Vitamin A	2000–2500 IU	Absence of deficiency signs[38]
Vitamin D		Not determined
Vitamin E	30 IU	Absence of deficiency signs[38]
Vitamin K		Not determined
Thiamine	10–15 mg	Maximum liver storage[38]
Riboflavin	20–25 mg	Maximum liver storage[38]
Pyridoxine	15–20 mg	Maximum liver storage[38]
Pantothenic acid	40–50 mg	Maximum liver storage[38]
Niacin	150–200 mg	Maximum liver storage[38]
Biotin	1–1.5 mg	Absence of deficiency signs[38]
Folic acid	6–10 mg	Maximum liver storage[38]
Vitamin B_{12}	0.015–0.02 mg	Maximum liver storage, absence of microcytic anemia[38]
Ascorbic acid	100–150 mg	Maximum tissue concentration[60]
Choline	600–800 mg	Growth, maximum liver storage[38]
myo-Inositol	300–400 mg	Growth, feed/gain ratio[38]

essential vitamins.[36-38] The reported requirements of Pacific salmon for specific vitamins are shown in Table 5. Halver[36-38] reported that the required dietary level for each vitamin was chosen based on salmon growth rate, feed conversion values, and maximum liver storage levels, but the actual data upon which the requirement was based were not reported. Baker[39] described the potential pitfalls of using either growth rates, feed conversion values, maximum enzyme activity, or maximum liver storage levels to establish dietary nutrient requirements. The actual requirement can vary considerably depending upon which criterion is used. Recent research with spring chinook salmon indicates that the dietary requirements for folic acid, pyridoxine, and riboflavin may be lower than previously reported.[40] Based on growth rates, dietary folic acid levels of 2 mg/kg dry diet and dietary pyridoxine levels of 6 mg/kg diet were sufficient in a purified diet formulation.

Based on both growth rates and maximum liver storage, dietary riboflavin levels of 7 mg/kg dry diet were sufficient. For pantothenic acid, a dietary level of 17 mg/kg dry diet supported maximum growth rates, while a dietary level of 214 mg/kg dry diet was required to achieve maximum liver storage. In light of this information, the vitamin requirements listed in Table 5 must be considered tentative. Further work is required to validate the required levels, which must be based on growth, feed efficiency, tissue storage levels, and appropriate enzyme activity determinations for those vitamins for which valid methodology exits. The findings of Leith et al.[40] must be confirmed before changes in recommended vitamin levels in Pacific salmon diets are justified. For practical manufacturing of Pacific salmon feed, the industry has been well served during the past 30 years using the levels shown in Table 5. Rarely is a vitamin deficiency encountered in Pacific salmon hatcheries or growout farms. However, as the information shown in Table 5 and described above illustrates, the actual vitamin requirements of Pacific salmon are not well studied under a variety of production conditions, fish sizes, or life stages, nor have the vitamin requirements of sockeye, pink, chum, or masu salmon been investigated.

MINERALS

Mineral requirements of fish are difficult to determine because of the problems associated with determining the contribution of dissolved minerals in rearing water toward the total needs

TABLE 6
Mineral Requirements of Pacific Salmon

Mineral	Chinook	Chum
Calcium		
Phosphorus (%)	R[61]	0.6[13]
Sodium		
Potassium (%)	0.8[62]	
Magnesium		
Copper		
Iron		
Manganese	R[61]	R[61]
Iodine (μg/kg diet)	0.6–1.1[15]	R[61]
Selenium	R[61]	R[61]
Zinc	R[61]	R[61]

[a] R indicates required.

of the fish. Pacific salmon are reared to the fry and fingerling stages in fresh water that has relatively low concentrations of dissolved minerals due to the fact that most Pacific salmon are found in coastal areas having very soft water. Rearing of postjuvenile Pacific salmon generally occurs in salt water, where the concentration of dissolved minerals is relatively high. Therefore, the dietary mineral requirements of Pacific salmon reared in fresh water should be different from those of Pacific salmon reared in salt water. However, experimental evidence to document the quantitative dietary mineral requirements of Pacific salmon in fresh water of various degrees of hardness and in salt water is presently lacking. Technical difficulties, lack of appropriate laboratory facilities, and the expense associated with rearing large Pacific salmon on purified diets all contribute to this lack of information (Table 6). An additional problem in determining the dietary mineral requirements of Pacific salmon is the manufacture of purified diets containing very low levels of certain essential minerals when casein is used as the principal protein source. A final difficulty is associated with the methods used to judge adequate dietary intake. Adequate dietary intake can be estimated by comparing growth rates, activity of specific enzymes, or the levels of a mineral in specific tissues or the whole body of the fish. It is not difficult to find examples in the literature of several different dietary requirements of a specific mineral for a species of fish determined using different methods of establishing adequate dietary intake. Often, fish fed diets deficient in essential minerals stop feeding and growing fairly soon after the beginning of a feeding trial, and this anorexia influences tissue or whole-body content of minerals by reducing the rate of dilution by growth.

Using values for dietary mineral requirements determined with purified or semipurified diets to formulate practical diets is a perilous practice. There are a number of specific instances where a dietary mineral level deemed adequate in a purified diet has proven to be inadequate in a practical diet, due to interaction among dietary constituents that results in reduced availability of the mineral and deficiency signs in fish. An example in Pacific salmon occurred in 1981, when an outbreak of irreversible cataracts in fish reared in North American freshwater hatcheries occurred, ultimately affecting over 40 million fish. Subsequent investigation suggested that the condition was the result of reduced dietary zinc bioavailability resulting from the use of high-ash fish meal and cottonseed meal (containing phytic acid). The diet contained approximately 65 mg zinc/kg, which was well above the reported requirement of 15 to 30 mg/kg determined in rainbow trout using purified diets.[41] However, examination of the whole-body levels of affected fish showed that the fish were deficient in zinc.[42] Fortification of the diet with zinc prevented subsequent episodes of cataracts associated with zinc deficiency in

Pacific salmon hatcheries. Researchers later showed that dietary zinc availability was reduced in fish fed diets containing elevated levels of calcium and phosphorus,[43] and elevated levels of calcium and phosphorus in combination with phytic acid.[44]

CAROTENOID PIGMENTS

Carotenoid pigments are not considered essential nutrients for Pacific salmon, but their presence in the diet is required to give the fish the flesh color normally associated with salmon, i.e., a deep pink-red. Astaxanthin and its esters are the principal carotenoid pigments found in wild Pacific salmon, and fish raised for market are fed diets containing astaxanthin and/or canthaxanthin, another carotenoid pigment that, like astaxanthin, occurs naturally in crustacea and other natural food items of Pacific salmon. Both astaxanthin and canthaxanthin are available as synthetic products that can be added to Pacific salmon diets. In addition, crustacean processing byproducts, dried *Phaffia* yeast, and certain algae products, all of which contain astaxanthin, are sometimes added to Pacific salmon diets to pigment the flesh. Broodstock Pacific salmon may require a small amount of dietary carotenoid pigment to ensure viability of the eggs, but this has not yet been adequately investigated. Torrissen et al.[45] recently reviewed carotenoid pigmentation in salmonids, and readers desiring more information than that presented here are suggested to begin with that review.

PRACTICAL DIET FORMULATION

Practical diets for Pacific salmon represent the range of possibilities for fish, excluding diet formulations designed for larval diets. The general types of diets are moist pellets, dry compressed pellets, and dry extruded pellets, and the general equipment and processing steps used to make each kind of pellet have been described by Hardy.[4] The standard for moist diets is the OMP, which has three formulations, as shown in Table 7. The mash is used as a starter feed, while the OP-2 and OP-4 formulations are designed for use after the fry are actively feeding and have reached a size at which they can consume 0.8 mm (1/32 in.) pellets. In general, Pacific salmon 32 mm (1.25 in.) long, which weigh approximately 0.35 g, can accept a pellet of this size. In practice, fry are often fed the mash until they reach a larger size to ensure continued growth. The OMP formulations all contain 28 to 32% moisture, and thus, must be kept frozen to prevent spoilage. In Japan, diets derived from the OMP are used to rear coho salmon in marine net pens. These diets differ from the OMP in that the wet portion of the diet, which is hydrolyzed and pasteurized whole Pacific whiting, or combinations of fish processing waste and whole fish, has been replaced with frozen sardines. The level of frozen fish is often higher than the fish hydrolysate used in North American diets. The frozen fish are partially thawed, ground, and combined with a dry mixture containing fish meal, wheat byproducts, vitamin and mineral premixes, and other dry ingredients, pelleted at the farm site, and used within 1 to 2 d. A similar process is used when fish hydrolysates, such as fish silage, are used in a moist salmon diet. Use of hydrolysates in Pacific salmon diets is confined to situations or locations where an inexpensive source of fish waste is available. At present, only a few Pacific salmon farms use diets containing fish hydrolysate, and they are all manufactured at or near the farm site.

Dry compressed diets, or conventionally pelleted diets, have replaced moist pellets in many hatcheries raising coho salmon juveniles for release, and in many private hatcheries raising fish for market. The principal reasons for this are that the cost of dry pellets is less than that for moist pellets, and the use of dry pellets eliminates the need for frozen storage of the feed at the hatchery or farm before use. The first dry-pellet formulation widely used in Pacific salmon rearing was developed at the Abernathy Salmon Culture Technology Center in Washington State.[47] This formulation has been modified since its development to reflect

TABLE 7
Oregon Moist Pellet Formulations

Ingredient	Oregon mash OM-3	Oregon pellet OP-4	Oregon pellet OP-2
Herring meal (or anchovy or hake meals up to 1.2 fish meals) except mash	49.9	47.5	28.0
Wheat germ meal	10.0	Remainder	Remainder
Cottonseed meal (48.5% CP)	—	—	10.0[a]
or poultry byproduct meal	—	—	8.0[a]
Dried whey	8.0	4.0	5.0
Corn distillers dried solids	—	—	4.0
Sodium bentonite	—	3.0	—
Vitamin premix[b]	1.5	1.5	1.5
Trace mineral premix[c]	0.1	0.1	0.1
Choline chloride (70%)	0.5	0.5	0.5
Wet fish hydrolysate (pasteurized)	20.0	30.0	30.0
Fish oil	10.0	6.5–7.0	6.0–6.75

[a] Values represent maximums, together cannot exceed 15% of diet.

[b] Supplies the following per kg diet: vitamin A (palmitate or acetate), 1654 IU; α-tocopheryl acetate, 503 IU; menadione sodium bisulfite, 18 mg; thiamine mononitrate, 46 mg; riboflavin, 53 mg; pyridoxine·HCl, 30.8 mg; d-calcium pantothenate, 115 mg; niacin, 222 mg; biotin, 0.6 mg; vitamin B_{12}, 0.06 mg; ascorbic acid, 893 mg; folic acid, 16.5 mg; and myo-inositol, 132 mg.

[c] Supplies the following per kg diet: Zn as $ZnSO_4$, 75 mg; Mn as $MnSO_4$, 20 mg; Cu as $CuSO_4$, 1.54 mg; and I as $KIO_3 \cdot C_2H_8N_2 \cdot 2HI$, 10 mg.

changes in ingredient availability and price, and to take into account new information on Pacific salmon nutrient requirements. The current Abernathy formulations for Pacific salmon are shown in Table 8. Another diet formulation that gives excellent results with Pacific salmon was developed in the West Vancouver Laboratory of the Department of Fisheries and Oceans[46] (Table 9). Compressed pellets are used to rear Pacific salmon to market size in North America, South America (Chile), Japan, and New Zealand. Many commercial feed manufacturers have developed proprietary feed formulations for compressed pellets for Pacific salmon that support excellent fish growth. These formulations are all derived from the Abernathy diet, but are changed somewhat to reflect ingredient prices and ingredient availability in various areas. The most notable modification found in commercial diet formulations is the substitution of a portion of the fish meal with soybean meal, cottonseed meal, canola meal, and/or poultry byproduct meal, which can provide savings, depending upon the price of fish meal.

Dry extruded pellets are the final type of practical diet used in Pacific salmon farming. The extruded dry pellet was first developed in fish for catfish, where a floating pellet was needed to permit farmers to gauge feed consumption of fish in large ponds. The extrusion process utilizes higher temperatures, moisture levels, and pressure than does the compressed pellet process, and these factors combine to cook the mixture, resulting in higher availability of the starch in the diet to the fish. This is an advantage in catfish farming, because the omnivorous catfish can tolerate higher levels of dietary carbohydrate than carnivorous fish, such as Pacific salmon. Extruded pellet technology was then introduced in Norway for manufacturing feed for Atlantic salmon, which were found to benefit from levels of dietary lipid above 20%. This level of lipid in the diet was nearly impossible to achieve using compressed pellets without greatly compromising pellet strength. However, extruded pellets, having a lower density than

TABLE 8
Abernathy Salmon Diet Formulations

Ingredient	Starter S8-2	Crumbles (18-2)	Pellets (19-2)
Herring meal	58	55	50
Dried whey	5	5	5
Blood flour (or meal)	10	10	10
Condensed milk solubles	3	3	3
or			
Poultry byproduct meal	1.5	1.5	1.5
Wheat germ meal	—	5	5
Wheat middlings mill-run	6.27	7.22	12.22
or shorts			
Vitamin premix[a]	1.5	1.5	1.5
Choline chloride (60%)	0.58	0.58	0.58
Ascorbic acid	0.1	0.1	0.1
Trace mineral mixture[b]	0.05	0.1	0.1
Lignin sulfonate pellet binder	2	2	2
Fish oil	12	9	9
or			
Soybean lecithin (max. 2%)			

[a] Supplies the following per kg diet: vitamin A (palmitate or acetate), 6600 IU; vitamin D_3, 441 IU; α-tocopheryl acetate, 502 IU; menadione sodium bisulfite, 28 mg; thiamine mononitrate, 47 mg; riboflavin, 53 mg; pyridoxine·HCl, 30.8 mg; d-calcium pantothenate, 115 mg; niacin, 220 mg; biotin, 0.6 mg; vitamin B_{12}, 0.06 mg; folic acid, 12.7 mg; and myo-inositol, 132 mg.

[b] Supplies the following per kg diet: Zn as $ZnSO_4$, 75 mg; Mn as $MnSO_4$, 20 mg; Cu as $CuSO_4$, 1.54 mg; and I as $KIO_3 \cdot C_2H_8N_2 \cdot 2HI$, 10 mg.

compressed pellets by virtue of being expanded in the manufacturing process, absorb higher levels of top-dressed fish and/or plant oil without becoming soft. The success of the extruded pellet in European Atlantic salmon farming has resulted in the last several years in the North and South American salmon feed industries developing the capability of manufacturing extruded pellets. A typical extruded diet formulation for Pacific salmon is shown in Table 10.

While the floating or slowly sinking properties of extruded pellets are desirable for Pacific salmon farmers, the need for Pacific salmon feed containing lipid levels above 20% is questionable. Pacific salmon fed these high levels of dietary lipid are found, at harvest, to contain high levels of lipid in the flesh, which results in an inferior product, except for smoking. The main advantage of extruded pellets for rearing Pacific salmon in marine net pens is that it is easier for the people feeding the fish to use extruded pellets because of their buoyancy. Properly trained feeders can, however, obtain the same growth rates and feed efficiency using compressed pellets, which typically cost 20% less than extruded pellets. Therefore, deciding which type of pellet to feed to postjuvenile Pacific salmon involves assessment of the skills of those responsible for feeding the fish.

There are several other types of diets used to rear Pacific salmon that deserve mention. They are semimoist diets that do not require frozen storage. One kind of diet that is commercially available in North America is pelleted using cold extrusion, like OMP, but lower moisture levels are used (about 18 to 22%) and a combination of preservation techniques is employed to prevent spoilage.[4] A second kind of diet, also commercially available in North America, involves similar preservation techniques but uses compressed pellet equipment to manufacture. Both types of diets are said to be more palatable to Pacific salmon than dry compressed or extruded diets, but systematic, controlled feeding trials to confirm this have not

TABLE 9
West Van 33 Diet Formulation[46]

Ingredient	g/kg diet as fed
Steam-dried herring meal (crude protein, 66.9%)	504.2
Poultry byproduct meal (crude protein, 63.1%)	73.5
Dried whey (crude protein, 14.8%)	76.9
Blood flour (crude protein, 81.8%)	50.2
Shrimp meal (crude protein, 35.4%)	31.5
Freeze-dried euphausids (crude protein, 67.4%)	20.6[a]
Wheat middlings (crude protein, 16.4%)	69.7
Vitamin supplement[b]	43.0
Mineral supplement[c]	20.4
Herring or salmon oil, antioxidant	84.5
Permapell (lignin sulfonate binder)	18.9
Ascorbic acid	1.9
Choline chloride (60%)	4.7

[a] Instead of freeze-dried euphausids, whole frozen euphasids have been used.

[b] The vitamin supplement supplied the following per kg diet: vitamin A acetate, 9455 IU; cholecalciferol, 2269 IU; DL-α-tocopheryl acetate, 567 IU; menadione as hetrazeen or MSBC, 24.8 mg; D-calcium pantothenate, 182.9 mg; pyridoxine·HCl 42.2 mg; riboflavin, 56.7 mg; niacin, 284 mg; folic acid, 18.9 mg; thiamine mononitrate, 38.4 mg; biotin, 2.84 mg; cyanocobalamin, 0.057 mg; and inositol, 378 mg.

[c] The mineral supplement supplied the following levels of minerals per kg of diet: Mn as $MnSO_4 \cdot H_2O$, 69.2 mg; Zn as $ZnSO_4 \cdot 7H_2O$, 28.4 mg; Co as $CoCl_2 \cdot 6H_2O$, 0.94 mg; Cu as $CuSO_4 \cdot 5H_2O$, 3.29 mg; Fe as $FeSO_4 \cdot 7H_2O$, 47.3 mg; I as KI, 5.1 mg; and Na as NaCl, 1951 mg.

been conducted. Hatchery staff confirm, however, that intermediate moisture diets do sometimes provide an advantage as starter feeds for Pacific salmon, depending upon species, water temperature, and the physical facilities used to start the fry on feed.

SPECIAL DIETS

Special diets for use with Pacific salmon involve those used to rear broodfish and those used to impart special characteristics to fish destined for market. In the case of broodstock Pacific salmon, few studies have been published. There has been documentation of the physical and chemical changes in the flesh and the fish and in the developing eggs,[48] and the effect of dietary modification on reproductive performance.[35,48] Addition of levels of trace mineral premix above that used in grower diets was not shown to increase the levels of these trace elements in the maturing fish or in the developing eggs, nor was this addition shown to influence reproductive performance of coho salmon.

Replacing fish oil with soybean oil or tallow was shown to modify the fatty acid composition of the neutral and polar lipid fractions of both maternal tissues and eggs in coho salmon, but this dietary change did not alter reproductive performance. Increasing the ascorbic acid content of the diet from 1045 to 10,450 mg/kg (430 to 2200 mg/kg ascorbic acid remained after pelleting and storage) did result in a significant increase in the ascorbic acid content of coho salmon eggs at spawning from 310 to 513 µg/g, but percent survival to hatch between the two groups was identical at 92%.[49] Increasing the dietary level of α-tocopheryl acetate by ten times over normal levels resulted in a decrease in percent survival to hatch of coho salmon eggs.

Increasing the dietary lipid content of broodstock coho salmon diets by adding an additional 4% herring oil did not influence egg hatchability, proximate composition of the eggs,

TABLE 10
Typical Salmon Extruded Feed Formulation

Ingredient	Percent
Fish meal	50
Poultry byproduct meal	8
Wheat flour	10.35
Whole wheat	5
Wheat mill-run	5
Soybean meal	5
Fish oil[a]	12
Molasses	2.5
Vitamin premix[b]	1
Choline chloride	0.5
Trace mineral premix[c]	0.1
Astaxanthin/canthaxanthin	0.05
Ascorbic acid	0.5

[a] Three to 4% fish oil is added before pelleting; the remainder is top-dressed after pelleting.

[b] Supplies the following per kg feed: vitamin A palmitate or acetate, 1654 IU; α-tocopherol acetate, 503 IU; menadione sodium bisulfite, 18 mg; thiamine mononitrate, 46 mg; riboflavin, 53 mg; pyridoxine·HCl, 38.6 mg; D-calcium pantothenate, 115 mg; niacin, 222 mg; biotin, 0.6 mg; vitamin B_{12}, 0.02 mg; ascorbic acid, 893 mg; myo-inositol, 132 mg; and folic acid, 16.5 mg.

[c] Supplies the following per kg feed: Zn and $ZnSO_4$, 75 mg; Mn as $MnSO_4$, 20 mg; Cu as $CuSO_4$, 1.54 mg; and I as KIO_3 or $C_2H_8N_2 \cdot 2HI$, 10 mg.

egg size, or egg polar or neutral lipid fatty acid composition.[49] The only modification of Pacific salmon grower diets shown to influence the reproductive performance of coho salmon to date is doubling of the vitamin premix level. Several years of feeding trials in commercial facilities have yielded promising, yet nonsignificant, increases in fecundity and egg hatchability in coho salmon fed diets containing double the level of vitamin premix compared to grower diets.[50] However, difficulties associated with conducting scientific feeding trials at commercial aquaculture facilities have continuously compromised the data. Comparison of activities of certain enzymes for which specific vitamins are essential cofactors have shown that, during the maturation process, the levels of activity are higher in fish fed diets containing higher levels of vitamins than in fish fed grower diets,[50] suggesting that further effort in this area is required.

The modification of grower diets to impart desirable characteristics to fish destined for market is an area that has received little attention, with the exception of carotenoid pigmentation.[46] Studies conducted with other salmonids suggest that this area of research could yield valuable information. For example, it has been demonstrated that modifying the diet by replacing herring oil with menhaden oil, which has a higher n-3 fatty acid content, increases the n-3 fatty acid content of Atlantic salmon after 6 months of feeding.[51] This finding in Atlantic salmon will most likely be applicable to Pacific salmon. Other areas of investigation include carotenoid pigmentation, which continues to be a problem in chinook salmon due to variability from fish to fish, modifying the total fat content or feeding level of fish destined for fresh sales or smoking, and modifying the diet to produce fish of consistent flavor. It is well known that diet influences the flavor of fish, and anectodal evidence suggests that improvements are required in farm-raised Pacific salmon.

A final area in which improvements are desirable is in the texture of the flesh. Whether this can be modified by altering the dietary lipid content, or whether this is found to be determined by environmental factors, such as exercise, has yet to be determined. Research with rainbow trout has shown that the frozen storage stability of the fish can be improved by feeding a diet high in α-tocopheryl acetate before harvest.[52] This improvement was proportional to the level of α-tocopherol found in the flesh of the trout, which was, in turn, directly proportional to the level of α-tocopheryl acetate added to the diet. Although growers of Pacific salmon primarily produce fish for fresh markets, this situation may change in the future, and dietary modification may therefore be desirable.

ACKNOWLEDGMENT

The author wishes to thank Dr. Robert R. Stickney for his valuable editorial assistance.

REFERENCES

1. **Craig, J. A. and Hacker, R. L.**, The History and Development of the Fisheries of the Columbia River, U.S. Bureau of Fisheries, 49, 133, 1940.
2. **Mahnken, C. V. M.**, personal communication, 1990.
3. **Embody, G. C. and Gordon, M.**, A comparative study of natural and artificial foods of brook trout, *Trans. Am. Fish. Soc.*, 54, 185, 1924.
4. **Hardy, R. W.**, Diet preparation, in *Fish Nutrition*, 2nd ed., Halver, J. E., Ed., Academic Press, New York, 1989, 475.
5. **Halver, J. E.**, Nutrition of salmonid fishes. III. Water-soluble vitamin requirements of chinook salmon, *J. Nutr.*, 62, 225, 1957.
6. **Wood, J. W.**, Diseases of Pacific Salmon — Their Prevention and Treatment, 3rd ed., State of Washington, Department of Fisheries, Olympia, WA, 1979.
7. **McLaren, B. A., Keller, E., O'Donnell, D. J., and Ellehjem, C. A.**, The nutrition of rainbow trout. I. Studies of vitamin requirements, *Arch. Biochem.*, 15, 169, 1947.
8. **Wolf, L. E.**, Diet experiments with trout, *Prog. Fish-Cult.*, 13, 17, 1951.
9. **Halver, J. E. and Coates, J. A.**, A vitamin test diet for long-term feeding studies, *Prog. Fish-Cult.*, 19, 112, 1957.
10. **National Research Council**, *Nutrient Requirements of Trout, Salmon, and Catfish*, National Academy of Sciences, Washington, D.C., 1973.
11. **Halver, J. E.**, Nutrition of salmonid fishes. IV. An amino acid test diet for chinook salmon, *J. Nutr.*, 62, 245, 1957.
12. **Ketola, H. G.**, Amino acid nutrition of fishes: requirements and supplementation of diets, *Comp. Biochem. Physiol.*, 73B, 17, 1982.
13. **Watanabe, T., Murakami, A., Takeuchi, L., Nose, T., and Ogino, C.**, Requirement of chum salmon held in freshwater for dietary phosphorus, *Bull. Jpn. Soc. Sci. Fish.*, 46, 361, 1980.
14. **Wekell, J. C., Shearer, K. D., and Houle, C. R.**, High zinc supplementation of trout diets, *Prog. Fish-Cult.*, 45, 144, 1983.
15. **Woodall, A. N., Ashley, L. M., Halver, J. E., Olcott, H. S., and Van Der Veen, J.**, Nutrition of salmonid fishes. XIII. The alpha-tocopherol requirements of chinook salmon, *J. Nutr.*, 84, 125, 1964.
16. **Hardy, R. W. and Shearer, K. D.**, unpublished data, 1987.
17. **Wilson, R. P.**, Amino acids and proteins, in *Fish Nutrition*, 2nd ed., Halver, J. E., Ed., Academic Press, New York, 1989, 111.
18. **Zeitoun, I. H., Halver, J. E., Ullrey, D. E., and Tack, P. I.**, Influence of salinity on protein requirements of rainbow trout (*Salmo gairdneri*) fingerlings, *J. Fish. Res. Board Can.*, 30, 1867, 1973.
19. **DeLong, D. C., Halver, J. E., and Mertz, E. T.**, Nutrition of salmonoid fishes. VI. Protein requirements of chinook salmon at two water temperatures, *J. Nutr.*, 65, 589, 1958.
20. **McCallum, I. M. and Higgs, D. A.**, An assessment of processing effects on the nutritive value of marine protein sources for the juvenile chinook salmon (*Oncorhynchus tshawytscha*), *Aquaculture*, 77, 181, 1989.
21. **Fowler, L. G.**, Substitution of soybean and cottonseed products for fish meal in diets fed to chinook and coho salmon, *Prog. Fish-Cult.*, 42, 86, 1980.

22. **Murai, T., Ogata, H., Kosutarak, P., and Arai, S.,** Effects of methanol treatment and amino acid supplementation of utilization of soy flour by chum salmon fingerlings, *Bull. Natl. Res. Inst. Aquaculture,* 12, 37, 1987.
23. **Arai, S., Yano, R., Deguchi, Y., and Nose, T.,** Essential and nonessential amino acids for growth of coho fingerling (*Oncorhynchus kisutch*), in *Proc. North Pacific Aquaculture Symp.,* Anchorage, AK, 1980, 285.
24. **Chance, R. D., Mertz, E. T., and Halver, J. E.,** Nutrition of salmonoid fishes. XII. Isoleucine, leucine, valine and phenylalanine requirements of chinook salmon and interrelations between isoleucine and leucine for growth, *J. Nutr.,* 83, 177, 1964.
25. **Halver, J. E. and Shanks, W. E.,** Nutrition of salmonid fishes. VIII. Indispensable amino acids for sockeye salmon, *J. Nutr.,* 72, 340, 1960.
26. **Akiyama, T., Murai, T., and Mori, K.,** Role of tryptophan metabolites in inhibition of spinal deformity of chum salmon fry caused by tryptophan deficiency, *Bull. Jpn. Soc. Sci. Fish.,* 52, 1255, 1986.
27. **Yu, T. C. and Sinnhuber, R. O.,** Effects of dietary n-3 and n-6 fatty acids on growth and feed conversion efficiency of coho salmon (*Oncorhynchus kisutch*), *Aquaculture,* 16, 31, 1979.
28. **Takeuchi, T., Watanabe, T., and Nose, T.,** Requirement for essential fatty acids of chum salmon (*Oncorhynchus keta*) in freshwater environment, *Bull. Jpn. Soc. Sci. Fish.,* 45, 1319, 1979.
29. **Donanjh, B. S., Higgs, D. A., Plotnikoff, M. D., Markert, J. R., and Buckley, J. T.,** Preliminary evaluation of canola oil, pork lard, and marine lipid singly and in combination as supplemental dietary lipid sources for juvenile fall chinook salmon (*Oncorhynchus tshawytscha*), *Aquaculture,* 68, 325, 1988.
30. **Hardy, R. W., Iwaoka, W. T., and Brannon, E. L.,** A new dry diet with alternative oil sources for Pacific salmon, in *Proc. World Maricul. Soc.,* Honolulu, Hawaii, 1979, 728.
31. **Mugrditchian, D. S., Hardy, R. W., and Iwaoka, W. T.,** Linseed oil and animal fat as alternative lipid sources in dry diets for chinook salmon (*Oncorhynchus tshawytscha*), *Aquaculture,* 25, 161, 1981.
32. **Yu, T. C. and Sinnhuber, R. O.,** Use of beef tallow as an energy source in coho salmon (*Oncorhynchus kisutch*) rations, *Can. J. Fish. Aquat. Sci.,* 38, 367, 1981.
33. **Hardy, R. W., Masumoto, T., Fairgrieve, W. T., and Stickney, R. R.,** The effects of dietary lipid source on muscle and egg fatty acid composition and reproductive performance of coho salmon (*Oncorhynchus kisutch*), in *The Current Status of Fish Nutrition in Aquaculture,* Takeda, M. and Watanabe, T., Eds., Tokyo University of Fisheries, Tokyo, 1990, 347.
34. **Buhler, D. R. and Halver, J. E.,** Nutrition of salmonoid fishes. IX. Carbohydrate requirements of chinook salmon, *J. Nutr.,* 74, 307, 1961.
35. **Coates, J. A. and Halver, J. E.,** Water-soluble vitamin requirements of silver salmon, U. S. Fish and Wildl. Serv., Spec. Sci. Rept., *Fisheries,* 281, 1, 1958.
36. **Halver, J. E.,** Nutrition in marine aquaculture, in *Marine Aquaculture,* McNeil, W. J., Ed., Oregon State University Press, Corvallis, OR, 1970, 75.
37. **Halver, J. E.,** The vitamins, in *Fish Nutrition,* Halver, J. E., Ed., Academic Press, New York, 1972, 29.
38. **Halver, J. E.,** The vitamins, in *Fish Nutrition,* 2nd ed., Halver, J. E., Ed., Academic Press, New York, 1989, 31.
39. **Baker, D. H.,** Problems and pitfalls in animal experiments designed to establish dietary requirements for essential nutrient, *J. Nutr.,* 116, 2339, 1986.
40. **Leith, D., Holmes, J., and Kaattari, S.,** Effects of Vitamin Nutrition on the Immune Response of Hatchery-Reared Salmonids, Final Report, Project 84-45A and 84-45B, Bonneville Power Administration, Portland, OR, 1990.
41. **Ogino, C. and Yang, G.-Y.,** Requirement of rainbow trout for dietary zinc, *Bull. Jpn. Soc. Sci. Fish.,* 44, 1015, 1978.
42. **Hardy, R. W. and Shearer, K. D.,** unpublished data, 1983.
43. **Hardy, R. W. and Shearer, K. D.,** Effect of dietary calcium, phosphate, and zinc supplementation on whole body zinc concentration of rainbow trout (*Salmo gairdneri*), *Can. J. Fish. Aquat. Sci.,* 42, 181, 1985.
44. **Richardson, N. L., Higgs, D. A., Beames, R. M., and McBride, J. R.,** Influence of dietary calcium, phosphorus, zinc, and sodium phytate level on cataract incidence, growth, and histopathology in juvenile chinook salmon (*Oncorhynchus tshawytscha*), *J. Nutr.,* 115, 553, 1985.
45. **Torrissen, O. J., Hardy, R. W., and Shearer, K. D.,** Pigmentation on salmonids — carotenoid deposition and metabolism, *Rev. Aquatic Sci.,* 1, 209, 1989.
46. **Higgs, D. A.,** personal communication, 1990.
47. **Fowler, L. G. and Burrows, R. E.,** The Abernathy salmon diet, *Prog. Fish-Cult.,* 33, 67, 1971.
48. **Hardy, R. W., Shearer, K. D., and King, I. B.,** Proximate and elemental composition of developing eggs of pen-reared coho salmon (*Oncorhynchus kisutch*) fed production and trace element-fortified diets, *Aquaculture,* 43, 147, 1984.
49. **Hardy, R. W. and King, I. B.,** unpublished data, 1984.
50. **Hardy, R. W. and Masumoto, T.,** unpublished data, 1986.

51. **Hardy, R. W., Scott, T. M., and Harrell, L. W.,** Replacement of herring oil with menhaden oil, soybean oil, or tallow in the diets of Atlantic salmon raised in marine net-pens, *Aquaculture*, 62, 267, 1987.

52. **Boggio, S. M., Hardy, R. W., Babbitt, J. K., and Brannon, E. L.,** The influence of dietary lipid source and alpha-tocopheryl acetate level on product quality of rainbow trout (*Salmo gairdneri*), *Aquaculture*, 51, 13, 1985.

53. **Klein, R. G. and Halver, J. E.,** Nutrition of salmonoid fishes: arginine and histidine requirements of chinook and coho salmon, *J. Nutr.*, 100, 1105, 1970.

54. **Akiyama, T., Arai, S., Murai, T., and Nose, T.,** Threonine, histidine, and lysine requirements of chum salmon fry, *Bull. Jpn. Soc. Sci. Fish.*, 51, 635, 1985.

55. **Halver, J. E., DeLong, D. C., and Mertz, E. T.,** Threonine and lysine requirements of chinook salmon, *Fed. Proc., Fed. Am. Soc. Exp. Biol.*, 17 (Abstr.), 1873, 1958.

56. **DeLong, D. C., Halver, J. E., and Mertz, E. T.,** Nutrition of salmonoid fishes. X. Quantitative threonine requirements of chinook salmon at two water temperatures, *J. Nutr.*, 76, 174, 1962.

57. **Halver, J. E., DeLong, D. C., and Mertz, E. T.,** Methionine and cystine requirements of chinook salmon, *Fed. Proc., Fed. Am. Soc. Exp. Biol.*, 18 (Abstr.), 2076, 1959.

58. **Halver, J. E.,** Tryptophan requirements of chinook, sockeye, and silver salmon, *Fed. Proc., Fed. Am. Soc. Exp. Biol.*, 24 (Abstr.), 169, 1965.

59. **Akiyama, T., Arai, S., Murai, T., and Nose, T.,** Tryptophan requirement of chum salmon fry, *Bull. Jpn. Soc. Sci. Fish.*, 51, 1005, 1985.

60. **Halver, J. E.,** Vitamin requirements, in *Fish Nutrition*, Neuhaus, O. and Halver, J. E., Eds., Academic Press, New York, 1969, 209.

61. **Lall, S. P.,** The minerals, in *Fish Nutrition*, 2nd ed., Halver, J. E., Ed., Academic Press, New York, 1989, 219.

62. **Shearer, K. D.,** Dietary potassium requirement of juvenile chinook salmon, *Aquaculture*, 73, 119, 1988.

PUFFER FISH, *FUGU RUBRIPES*

Akio Kanazawa

INTRODUCTION

The puffer fish has the highest market value of any fish for "Fugu cooking" in Japan, Korea, and China, in spite of the fact that their internal organs contain a potent poison called tetrodotoxin. Since the artificial seed production of puffer fish was successfully developed in 1962, it has been cultured from the larval stage to marketable size in Japan. About 7,200,000 puffer fish larvae of 50 mm in length were produced for aquaculture and released to the sea in 1987. In Japan, the total production of puffer fish was only 69 t in 1980, but it increased to 1028 t in 1987.

NUTRIENT REQUIREMENTS

TEST DIET DEVELOPMENT

Feeding trials were conducted to study the effect of levels of dietary protein (casein), lipid (pollack liver oil), minerals,[1] and vitamins on the growth of 2.0 g puffer fish. Growth response was examined in terms of weight gain, feed conversion efficiency, protein efficiency ratio, and net protein utilization in a 3-week feeding experiment at 25 to 26°C.[2]

Results from these studies indicated that the optimum dietary protein level was about 50% when casein was used as the protein source (Figure 1). Also, the optimum level of lipid, minerals, and vitamins were estimated to be about 6 (or less), 4, and 3% of the diet, respectively (Figure 2).

PROTEINS

Four test diets containing various protein sources to simulate the amino acid profile of the body protein of larval puffer fish were formulated using a computer (Table 1). Feeding experiments were conducted using larval puffer fish, 34 d of age and 8.05 mm in length. Microparticulate diets with particle sizes of 350 to 500 μm were used. The results of the feeding trial are shown in Figure 3. Growth of larvae fed diets containing 37% brown fish meal or 37% squid meal was superior to that of larvae fed the diet containing 31% soybean protein.[3]

CARBOHYDRATES

Basic information on the development of an efficient compound diet for culturing puffer fish is necessary. Five test diets containing 50% white fish meal were prepared, each of which contained different levels of dextrin, feed oil, and vitamins, and were fed to puffer fish weighing 28.6 g for 80 d. Growth of the fish fed the test diets were slightly lower than that of the control group fed raw sand lance, *Ammodytes personatus*. No specific abnormalities were observed in the lateral muscle and the digestive tract of either the control or the test diet groups. The optimum dietary dextrin level was found to be approximately 30%. Supplementation of 10% pollack liver oil slightly improved growth. However, doubling the vitamin content (6%) showed no significant effect on growth.[4]

FATTY ACIDS

The bioconversion of [1-^{14}C]linoleic acid (18:2n-6)[5] and [1-^{14}C] linolenic acid (18:3n-3)[6] to highly unsaturated fatty acids has been investigated in puffer fish. After injection of the radioactive precursors, the proportional radioactivity was measured in each fatty acid, which

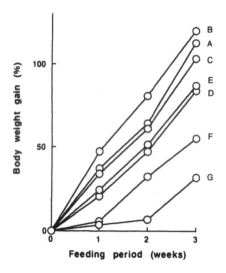

FIGURE 1. Effect of protein levels on body weight gain of puffer fish. A: 60%; B: 50%; C: 45%; D: 40%; E: 30%; F: 20%; G: 0% protein. (From Kanazawa, A. et al., *Nippon Suisan Gakkaishi,* 46, 1359, 1980. With permission.)

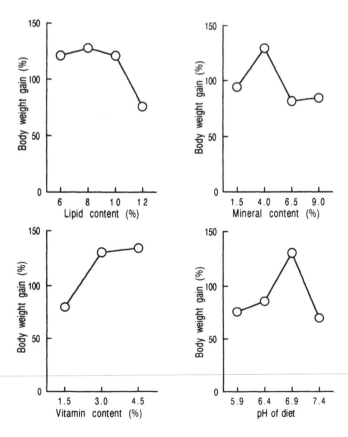

FIGURE 2. Effect of lipid, mineral, and vitamin levels as well as pH of diet on body weight gain of puffer fish. (From Kanazawa, A. et al., *Nippon Suisan Gakkaishi,* 46, 1360, 1980. With permission.)

TABLE 1
Composition of Test Diets for Puffer Fish
Expressed as g/100 g Dry Diet[3]

Ingredient	Diet 1	Diet 2	Diet 3	Diet 4
Brown fish meal	37			8
Squid meal		37		13
Soybean protein			31	13
White fish meal	9	7	8	8
Krill meal	9	5	8	8
Casein	9	8	8	8
Arginine·HCl	1			
Cellulose		7	9	6
Dextrin	7	8	8	8
Mineral mixture	5	5	5	5
Vitamin mixture	6	6	6	6
Soybean lecithin	4	4	4	4
Pollack liver oil	3	3	3	3
n-3 HUFA	2	2	2	2
Zein	8	8	8	8
Protein content (%)	49.74	50.52	50.45	50.41

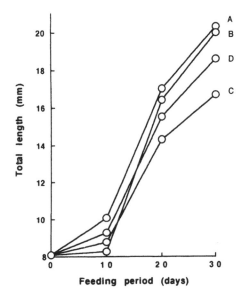

FIGURE 3. Effect of protein source on total length
of puffer fish. A: Squid meal; B: brown fish meal; C:
soybean protein; D: mixture.

was isolated by thin-layer chromatography on $AgNO_3$-Kieselgel G and preparative gas-liquid
chromatography. The puffer fish converted some 18:2n-6 to eicosadienoic acid (20:2n-6) and
arachidonic acid (20:4n-6), but there was little incorporation of labeled 18:3n-3 into
eicosapentaenoic acid (20:5n-3) and docosahexaenoic acid (22:6n-3).

The essential fatty acid requirement of puffer fish has also been demonstrated in feeding
experiments (Figure 4). Dietary 18:2n-6, 18:3n-3, and 20:4n-6 were not effective as dietary
essential fatty acid sources, whereas 20:5n-3 was the most effective dietary essential fatty
acid.[7]

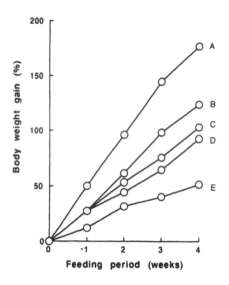

FIGURE 4. Effect of essential fatty acids on growth of puffer fish. A: 7% 18:1n-9 + 1% 20:5n-3; B: 7% 18:1n-9 + 1% 20:4n-6; C: 7% 18:1n-9 + 1% 18:3n-3; D: 7% 18:1n-9 + 1% 18:2n-6; E: 8% 18:1n-9. (From Kanazawa, A., in *Nutrition and Feeding in Fish*, Cowey, C. B. et al., Eds., Academic Press, London, 1985, 285. With permission.)

PHOSPHOLIPIDS

Some dietary phospholipids have been demonstrated to be indispensable for sustaining growth and survival of fish, particularly at the larval stage.[7,8] Feeding trials using 6 g puffer fish fingerlings were conducted to examine the effects of some dietary phospholipids on growth and survival.[9] A deficiency in dietary phospholipids caused total mortality 50 d after feeding. Chicken egg lecithin showed a higher nutritive value than soybean lecithin or rapeseed lecithin. Soybean lecithin in the diet at the 20% level resulted in a higher growth rate than at either a 1 or 1.5% level.

PRACTICAL DIET FORMULATION

Puffer fish have generally been reared by feeding raw minced trash fish, such as sand lance, sardines, or mackerel and krill. Recently, moist, dry and expanded pellets have been developed for puffer fish. Among these pellet types, the moist pellet is primarily used in Japan. In 1989, the annual production of artificial feed (as a powdered premix) for puffer fish was 2500 t. This premix is combined with trash fish to form the moist pellets prior to being fed.

The growth response of puffer fish fed a dry pellet diet, a moist pellet diet, and raw sand lance is shown in Figure 5.[10] The composition of a dry pellet diet for puffer fish is presented in Table 2.

MICROPARTICULATE DIET

Attempts to rear puffer fish larvae with microparticulate diets have been carried out in Yamaguchi Prefecture Inland Sea Fish Farming Center.[12] Three types of microparticulate

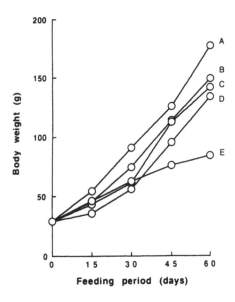

FIGURE 5. Growth of puffer fish fed various diets. A: Sand lance fed group; B: moist pellet fed group; C: krill fed group; D: dried pellet fed group; E: sardine fed group. (From Namba, K. et al., *Suisanzoshoku,* 36, 55, 1988. With permission.)

TABLE 2
Composition of a Dry Pellet Diet for Puffer Fish[11]

Ingredient	Percent
White fish meal	50
Dextrin	25
Cellulose	4
Pollack liver oil	5
Vitamin mixture	3
Mineral mixture	12
Binder	1

From Nakagawa, H., *Fish Culture,* 27, 113, 1990. With permission.

diets (A, B, and C) were prepared by Kyowa Hakko Kogyo Co., Ltd. Twenty-three day-old larvae, 5.4 mm in length, were reared for 37 d on a combination of microparticulate diets and live food. The feeding schedule of the two groups is shown in Figure 6. The results of the feeding experiment are presented in Figure 7. The puffer fish larvae readily consumed the combination of the microparticulate diets and live food. The results indicate that higher growth and survival rates were obtained in the group fed a combination of artificial and live food than the control group, which only received the live food.

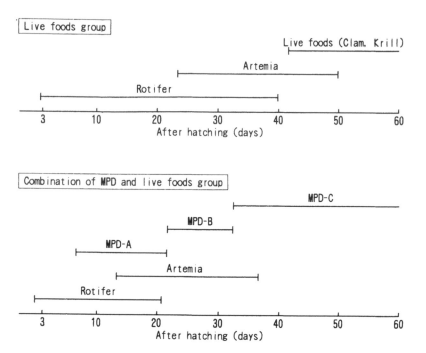

FIGURE 6. Feeding schedule used in the production of puffer fish larvae. MPD: microparticulate diet.

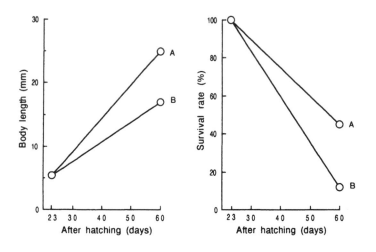

FIGURE 7. Growth and survival rates of puffer fish fed microparticulate diets and live food. A: Microparticulate diets + live food; B: live food only.

REFERENCES

1. **Cowey, C. B., Pope, J. A., Adron, J. W., and Blair, A.,** Studies on the nutrition of marine flatfish. The protein requirement of plaice (*Pleuronectes platessa*), *Br. J. Nutr.*, 28, 447, 1972.
2. **Kanazawa, A., Teshima, S., Sakamoto, M., and Shinomiya, A.,** Nutritional requirements of the puffer fish: purified test diet and the optimum protein level, *Nippon Suisan Gakkaishi*, 46, 1357, 1980.
3. **Kanazawa, A.,** unpublished data, 1990.
4. **Kakuta, I., Okabe, M., Nanba, K., Nakagawa, H., Kumai, H., and Nakamura, M.,** Studies on nutritional requirement of Fugu. III. Effect of dietary dextrin, feed oil and vitamins, *Suisanzoshoku*, 36, 183, 1988.
5. **Kanazawa, A., Teshima, S., and Imai, K.,** Biosynthesis of fatty acids in *Tilapia zillii* and the puffer fish, *Mem. Fac. Fish., Kagoshima Univ.*, 29, 313, 1980.
6. **Kanazawa, A., Teshima, S., and Ono, K.,** Relationship between essential fatty acid requirements of aquatic animals and the capacity for bioconversion of linolenic acid to highly unsaturated fatty acids, *Comp. Biochem. Physiol.*, 63B, 295, 1979.
7. **Kanazawa, A.,** Essential fatty acid and lipid requirements of fish, in *Nutrition and Feeding in Fish*, Cowey, C. B., Mackie, A. M., and Bell, J. G., Eds., Academic Press, London, 1985, 281.
8. **Kanazawa, A., Teshima, S., and Sakamoto, M.,** Effect of dietary bonito-egg phospholipids and some phospholipids on growth and survival of the larval ayu, *Plecoglossus altivelis*, *Z. Angew. Ichthyol.*, 4, 165, 1985.
9. **Kanazawa, A.,** unpublished data, 1990.
10. **Namba, K., Nakagawa, H., Okabe, M., Kakuta, I., Kumai, H., Nakamura, M., and Kasahara, S.,** Studie on nutritional requirements of Fugu. II. Histological changes caused by various diets, *Suisanzoshoku*, 36, 53. 1988.
11. **Nakagawa, H.,** Status and problem of nutritional research on puffer fish, *Fish Culture*, 27, 113, 1990.
12. Yamaguchi Prefecture Inland Sea Fish Farming Center, unpublished data, 1989.

RAINBOW TROUT, *ONCORHYNCHUS MYKISS*

C. Young Cho and Colin B. Cowey

INTRODUCTION

Rainbow trout is one of about 70 species of salmonids that occur in the natural environment. They are found in most of the cold waters of the north temperate zone, including Africa, north of the Atlas mountains, northwestern Mexico, and Taiwan. The salmonid family comprise Pacific salmon (*Oncorhynchus* sp.), Atlantic salmon, trout (*Salmo* sp.), char (*Salvelinus* sp.), graylings (*Thymallus* sp.), whitefish (*Coregonus* sp.), and several other genera.

Rainbow trout were formerly assigned to the genus *Salmo* but it has recently been established that they more properly belong to the genus *Oncorhynchus*.

In the natural environment rainbow trout feed on various invertebrates. Fry consume plankton but, as they increase in size, there is a shift in food pattern to insects and crustaceans, and thence to fish, if available. Rainbow trout will exist in a temperature range of approximately 1 to 20°C. They have an upper critical range of 19 to 30°C, with an upper incipient lethal temperature of 26°C.[1] A generalized optimum, with respect to growth rate and food conversion rate, occurs at approximately 15°C, and this has been referred to as the standard environmental temperature for rainbow trout.

Rainbow trout are cultured at high density in raceways, ponds, net cages, or even large tanks. They require a high throughput of clean water, largely because of their high oxygen demand. Individual trout farms may vary in size from those producing little more than 50 t/ annum to those producing a few thousand t/annum. At the present time almost all rainbow trout are raised on dry pelleted diets, the use of moist feeds made up largely of unwanted (so-called 'trash') fish being now regarded as environmentally unacceptable in many countries. The output of the principal trout-producing countries is shown in Table 1.

Eggs, stripped manually from mature females or forced from them by injecting air into the peritoneal cavity, are collected in a bucket or similar receptacle, and fertilized directly with milt obtained at the same time from male broodstock. The fertilized eggs are laid on incubation trays and incubated in running water at temperatures of 5 to 8°C in a hatchery. They develop through the eyed egg stage and hatch as yolk sac fry after about 370 degree days at 8°C; with further development, as the yolk sac is resorbed, swim-up fry emerge after 150 degree days at 8°C.[7] Swim-up fry are transferred to nursery tanks, where they are fed starter granules at frequent intervals. They grow rapidly as the water temperature is increased and will be graded and moved to larger growout facilities as appropriate.

Coldwater fish have very restricted, seasonal spawning periods; fish farmers, however, need stock throughout the year. Solutions to this problem have been sought either through breeding programs or by manipulation of spawning time through photoperiod control. There are no exact data on the former approach — genetic selection — but in northern latitudes there are now available several strains of rainbow trout that spawn at different times over the period September to April. Purdom[8] comments "whether these are representative of geographic race divergence or differentiation under domestication, is not known but it seems probable that both influences are responsible for the range of spawning times which is observed in domesticated strains of rainbow trout."

Photoperiod control is, by comparison with selection, quick; it appears to be effective, although the question of egg quality has not yet been entirely resolved. Appropriate equipment, which may be expensive, is necessary if photoperiod control is to be applied.

A major advance in the culture of rainbow trout has been the introduction of monosex (usually all female) stocks — a development that allows the possibility of going on to produce

TABLE 1
Annual Production of Rainbow Trout in Principal Trout-Producing Countries in the Northern Hemisphere

Country	Production (tons)	Year	Comments	Ref.
France	29,000	1988	Pan sized	2
Italy	29,000	1988	Pan sized	2
West Germany	15,000	1988	Pan sized	2
Spain	15,000	1988	Pan sized	2
Denmark	20,000	1988	Pan sized	3
	4,000		Large	
Finland	16,000	1988	Large	3
Norway	9,000	1988	Large	3
Sweden	7,500	1988	Large	3
U.S.A.	25,000	1988	Includes other trout species	4
Canada	4,000	1988	Includes other trout species	5
Japan	20,000	1987		6

sterile fish. Sexual development among male trout is particularly troublesome for the industry in that a sizable proportion of males become sexually mature before they reach marketable size. Apart from the attendant loss in food conversion efficiency and growth rate, the mature males have poor flesh quality and an unattractive appearance. Moreover, maturing males cannot make the necessary physiological adjustments to salt water and so cannot be cultured in seawater.

Sexual maturity in females is less of a problem in that it rarely occurs in fish under 2 years of age. However, an increasing market for larger fish (either for the table or for sport fishing), coupled with a tendency to grow such fish in seawater where spawning is frequently fatal, suggests a need for control of sexual maturation in females also.

The sex determining system in trout is an XX (female), XY (male) system.[9] This provided a means of making all female stocks; female fry could be masculinized by treating them with androgens before they reached sexual differentiation — this was achieved comparatively simply by giving first-feeding fry a diet containing 17 α-methyl testosterone for 600 to 700 degree days.[10] At maturity these masculinized females produced homogametic sperm, which, when used to fertilize eggs from normal females, would give rise only to female offspring (Y chromosomes being absent from the sperm). All female stocks obtained by this method began to be used in Europe in the early 1980s and are now in widespread use there.

Complete suppression of gonad development was observed in female triploid rainbow trout that had arisen spontaneously in hatchery stocks.[11] As triploidy could be induced in the eggs of rainbow trout by administration of a thermal shock,[12] the availability of all female stock has made the production of sterile stocks a practical possibility. Bye and Lincoln[10] showed that eggs held at 10°C, heat shocked 40 min after fertilization by immersion in water at 28°C for 10 min, had high levels of triploidy and acceptable levels of survival. Pressure treatment of fertilized eggs is another means of inducing triploids. Ovaries from 5-month-old triploid trout, while having a typical lamellar structure, do not contain oocytes. It is not yet clear whether sterile, triploid fish are likely to become a significant factor in commercial trout production.

LABORATORY CULTURE

The principal requirement for successful culture of rainbow trout is a plentiful supply of clean running water. The pH of the water should be in the range 6.7 to 8.0, it should be 80

TABLE 2
Water Requirement and Biomass Loading
in Experimental Tanks[13]

CONDITIONS:

Biomass (BMW), kg/l	0.05–0.07
Water exchange rate per hour (EXH)	2–3
Water volume in tank (WVT), l	*

FORMULAE:

Water flow rate (WFR), l/min
  = (WVT x EXH)/60 min

Water requirement per kg biomass (l/min per kg)
  = EXH/(BMW x 60 min)

Biomass loading per unit flow (kg/l/min)
  = (BMW/EXH) x 60 min

EXAMPLE:

Water volume in tank (WVT), l	65
Water flow rate (WFR), l/min	2.17–3.25
Water requirement, l/min/kg	0.48–0.71
Biomass loading, kg/l/min	1.40–2.10

Optimum water temperature(s):	15°C (13–17°C)
Oxygen requirement: approx 300 mg/kg body weight/h at 15°C	

* Will vary with individual aquarium system, design of tank will affect both the BMW and EXH chosen.

to 100% saturated with dissolved oxygen and should certainly contain not less than 5.0 mg dissolved oxygen/l when leaving the fish tanks. Alkalinity in the range 20 to 200 ppm (as calcium carbonate) is acceptable and suspended solids should be less than 80 ppm. Particular attention should be paid to chlorine levels, especially if domestic water is being used. These must not exceed 0.003 ppm; domestic water may be dechlorinated by passage through filters containing active charcoal. At or near neutral pH most of the ammonia is present as NH^+_4. The pK for the reaction $NH_3 + H^+ = NH^+_4$ being 9.58; NH^+_4 is not toxic, nevertheless ammonia levels are best maintained at not more than 0.01 ppm. Guidelines for estimating the water requirements necessary to support a given biomass of trout are shown in Table 2.

One of the dangers in laboratory culture systems is that of supersaturation leading to embolisms and mortalities among the fish. This may happen when, for example, well water at low temperature (3 to 5°C) is raised to an experimental temperature of say 15°C. Provision for aerating the water vigorously, such as a cascade or diffuser stones, is therefore necessary.

Water is often a limited commodity and water reuse systems that incorporate a biological filter have proved very useful under these circumstances. About 80 to 90% of the water in such systems is recirculated, with 10 to 20% of new water being introduced continuously. Self-cleaning tanks are another useful feature of laboratory aquarium systems; these may have a tank bottom sloping to one side of the rectangular tank from which drainage occurs. Alternatively, cylindrical tanks may have a gently sloping cone-shaped bottom with drainage through an outside standpipe.

Nutrient test diets are composed of purified components. For fish they are rarely, if ever, completely defined. The content of the nutrient under study should, however, be precisely established. The test diet shown in Table 3 has been used in this laboratory for studies on protein, vitamin, mineral, and essential fatty acid requirements of rainbow trout. It is based on casein and gelatin, and allows positive control of all these nutrients. For studies on minerals, it is necessary to define mineral concentrations in the water, as fish have a considerable capacity to take ions up from the water. Experiments are generally carried out on young, rapidly growing fish where demand for essential nutrients is likely to be at its greatest. The

TABLE 3
Semipurified Reference Diet for Salmonids

Ingredient	Percent
Casein, high nitrogen[a]	40
Gelatin	4
Starch, raw	11.5
Dextrin, white, technical	9
D-glucose	5
α-cellulose	3
Carboxymethyl cellulose[b]	—
DL-methionine	0.5
L-arginine	1
Vitamin premix[c]	3
Mineral premix[d]	8
Fish oil	15
Composition:	
Digestible energy (DE), MJ/kg feed	17
Digestible protein, g/MJ DE	22
Digestible lipid, g/kg	150
Dry matter digestibility coefficient, %	80
Protein digestibility coefficient, %	98
Energy digestibility coefficient, %	90

[a]   Use vitamin-free casein for vitamin experiments.
[b]   May be needed instead of α-cellulose when a moist diet is being fed
    by syringe.
[c]   See Table 5 for composition.
[d]   See Table 6 for composition.

diet may be steam pelleted in a laboratory pellet mill and subsequently crumbled to granules that are then sieved to appropriate size for the fish under experiment.

For studies on essential amino acid requirements, it is necessary to be able to provide a series of diets containing eight or so graded levels of the amino acid under test. This is best done by supplying half of the dietary protein as intact protein and half as a mixture of free amino acids. The content of the amino acid under test in the latter mixture can then be varied to provide appropriate levels in the eight or so diets needed. Unfortunately, diets containing large quantities of crystalline amino acids give rates of growth that are lower, often substantially lower, than those obtained when the diet contains whole protein.[14] To overcome this difficulty, test diets have been formulated in this laboratory in which the free amino acid component is coated with a solution of agar in water (the agar amounting to 1% of the diet) prior to being mixed with other components. The composition of this diet is shown in Tables 4 to 6. Rapid rates of growth have been obtained in this laboratory in experiments designed to measure the arginine requirement of rainbow trout fry.

NUTRIENT REQUIREMENTS

PROTEIN AND AMINO ACIDS

The protein requirement of rainbow trout was determined as 40 to 45% of the dry matter in the diet from experiments on juvenile fish.[15-17] In current practical diets, levels of digestible protein range from 33 to 42% depending on the energy density. The levels of digestible energy in these diets are in the range 15 to 17 MJ/kg and dietary protein comprises 22 to 25 g/MJ digestible energy.

Quantitative estimates of essential amino acid requirement by dose-response curve have

TABLE 4
Semipurified Test Diet with Low Concentrations of Arginine and Lysine

Ingredient	Percent
Corn gluten meal	15
Skim milk	25
Starch	14
α-cellulose	3
Amino acid premix (with agar, 1% of diet)	19
Vitamin premix[a]	3
Mineral premix[b]	6
Fish oil	15
Composition:	
Digestible energy (DE), MJ/kg feed	17
Digestible protein, g/MJ DE	19
Digestible lipid, g/kg	153
Total phosphorus, g/kg	4
Arginine, g/kg	6
Lysine, g/kg	8
Dry matter digestibility coefficient, %	75
Protein digestibility coefficient, %	98
Energy digestibility coefficient, %	85

[a] See Table 5 for composition.
[b] See Table 6 for composition.

TABLE 5
Vitamin Premix for Semipurified Diets

Vitamin	g/kg premix
Vitamin A (retinyl acetate)	233×10^3 IU
Vitamin D_3 (cholecalciferol)	1×10^5 IU
Vitamin E (dl-α-tocopheryl-acetate)	6.7
Vitamin K (menadione sodium bisulfite)	1.7
Ascorbic acid	13.3
Vitamin B_{12} (cyanocobalamin)	0.007
d-Biotin	0.013
Choline dihydrogen citrate	368
Folic acid	0.7
Inositol	17
Niacin	10
Ca-pantothenate	1.3
Pyridoxine·HCl	0.3
Riboflavin	0.3
Thiamine·HCl	1.3
Carrier (α-cellulose or starch)	Remainder

been made for only five amino acids (Table 7). For some of them (e.g., arginine) there is considerable disparity between the requirement values obtained in different laboratories. There is no immediate explanation for this interlaboratory variation, although factors such as differences in dietary protein level, amino acid sources, feed intake, and other aspects of methodology may have played a part. Consequently, value judgments were applied to arrive at the recommended figures given in Table 8.

Ogino[33] measured retention of individual amino acids in whole-body protein of rainbow

TABLE 6
Mineral Premix for Semipurified Diets

Mineral	g/kg premix
CaHPO$_4$ · 2H$_2$O (23% Ca, 18% P)	375
CaCO$_3$ (40% Ca)	37.5
NaCl (39% Na)	188
K$_2$SO$_4$ (45% K)	250
MgSO$_4$ (20% Mg)	125
FeSO$_4$ · 7H$_2$O (21% Fe)	88
MnSO$_4$ · H$_2$O (33% Mn)	3.8
ZnSO$_4$ · H$_2$O (36% Zn)	6.9
CuSO$_4$ · 5H$_2$O (25% Cu)	2.0
CoCl$_2$ · 6H$_2$O (25% Co)	0.33
KI (76% I)	0.2
Na$_2$SeO$_3$ (42% Se)	0.03
Carrier (α-cellulose)	Remainder

TABLE 7
Amino Acid Requirements of Juvenile Rainbow Trout

	Crude Protein in diet (%)	Requirement			
Amino acid		As % of dietary protein	As % of dry diet	Type of diet	Ref.
Arginine	36	3.3	1.2	Semipurified	18
	45	3.6	1.6	Semipurified	19
	35	4.0	1.4	Purified	20
	33	4.7	1.6	Semipurified	21
	47	5.9	2.8	Semipurified	22
Leucine	38	9.1	3.4	Semipurified	23
Lysine	35	3.7	1.3	Purified	24
	45	4.2	1.9	Semipurified	25
	47	6.1	2.9	Semipurified	22
Methionine	46.4	2.2[a]	1.0[a]	Purified	26
	35	3.0[b]	1.1[b]	Purified	27
	35	2.9[c]	1.0[c]	Purified	28
Tryptophan	55	0.5	0.3	Semipurified	29
	35	0.6	0.2	Purified	30
	42	1.4	0.6	Purified	31

[a] Diet lacking cystine.
[b] Diet contained 0.3% cystine.
[c] Diet contained 0.5% cystine.

trout and used the increase in essential amino acid content to estimate requirement. The method appears to assume that maintenance requirements in young growing trout are very low (although it is not easy to reconcile this view with the fact that only 30 to 40% of dietary N is retained in the growth of the fish), so that the pattern of amino acids deposited in body weight gain is the main factor determining patterns of amino acids required.

Rainbow trout are carnivorous and so have a high capacity to digest dietary protein. This is reflected in the high values obtained for apparent digestibility coefficients of proteinaceous materials (Table 9) obtained via the Guelph system.[34] Low values were encountered with

TABLE 8
Recommended Nutrient Concentrations in Production Diets for Rainbow Trout as Units per Kilogram Diet[32]

Energy concentration (digestible energy)	15 MJ
Protein: 22–25 g digestible protein/MJ digestible energy; equivalent to 35–38% digestible protein or approximately 40% crude protein	
Amino acids (based on a 34% crude protein diet)	
Arginine	1.5 g
Leucine	1.4 g
Lysine	1.8 g
Methionine (in presence of cystine)	1.0 g
Tryptophan	0.2 g
Essential fatty acid	
18:3n-3[a]	8–17 g
Vitamins	
Vitamin C	50 mg
Choline	3000 mg
Inositol	200 mg
Biotin	0.15 mg
Folates	5 mg
Niacin	10 mg
Pantothenic acid	12 mg
Riboflavin	4 mg
Thiamine	1.5 mg
Vitamin B_6	3.0 mg
Vitamin B_{12}	0.015 mg
Vitamin A	2000 IU
Vitamin D	2400 IU
Vitamin E	50 mg
Vitamin K	10 mg
Minerals	
Calcium	ND[b]
Phosphorus	6 g
Sodium	ND
Potassium	ND
Magnesium	0.5 g
Manganese	20 mg
Zinc	30 mg
Iron	60 mg
Copper	5 mg
Iodine	ND
Selenium	0.3 mg

[a] Requirement may vary with dietary energy density or dietary lipid level being about 20% of dietary lipid. The EFA requirement may be met by smaller amounts of 20:5n-3 and 22:6n-3. A small but undefined requirement for 18:2n-6 may exist.

[b] Not accurately determined.

certain refractory proteins, such as collagens or feather meal, which contain a high proportion of disulfide bonds, as well as in proteins that had been heat damaged during preparation.

FATTY ACIDS

The experimental evidence indicates that essential fatty acid (EFA) requirements of trout can be met entirely by a dietary supply of n-3 fatty acids (Table 8). Watanabe et al.[35] showed that 0.8% of 18:3n-3 was required for optimal growth; this was later modified to 20% of dietary lipid as 18:3n-3 or 10% of dietary lipid as 20:5n-3 or 22:6n-3.[36]

TABLE 9
Apparent Digestibility Coefficients and Digestible Energy Values[a] of Ingredients Measured with Rainbow Trout[40]

Ingredient	International feed no.	Digest. coeff. D.M. %	C.P. %	Lipid %	G.E. %	D.E. (MJ/kg)
Alfalfa meal	1-00-023	39	87	71	43	7.7
Blood meal, animal						
Spray dried	5-00-381	91	99	—	89	19.4
Flame dried	5-00-381	55	16	—	50	10.9
Brewer's dried yeast	7-05-527	76	91	—	77	13.9
Corn, yellow	4-02-935	23	95	—	39	6.6
Corn gluten feed	5-02-903	23	92	—	29	5.4
Corn gluten meal	5-09-318	80	96	—	83	17.6
Corn dist. dried sol.	5-02-844	46	85	71	51	10.7
Feather meal, poultry	5-03-795	75	58	—	70	15.7
Fish meal, herring	5-02-000	85	92	97	91	18.8
Meat & bone meal	5-09-321	78	85	73	85	15.0
Poultry byprod. meal	5-03-798	52	68	79	71	13.9
Rapeseed meal	5-03-871	35	77	—	45	8.1
Soybean, full fat,						
cook.	5-04-597	78	96	94	85	19.0
Soybean meal, dehull.	5-04-612	74	96	—	75	13.5
Wheat middlings	4-05-205	35	92	—	46	7.6
Whey, dehydrated	4-01-182	97	96	—	94	13.4
Starch, raw		37	—	—	49	8.6
Starch, gelatinized		86	—	—	86	13.3
Fish protein						
concentrate		90	95	—	94	17.2
Soybean protein						
concentrate		77	97	—	84	15.4
C201-Guelph						
Reference diet		71	91	92	79	16.2

[a] All values are cumulative averages of compiled data. D.M. = dry matter; C.P. = crude protein; G.E. = gross energy; D.E. = digestible energy; 1 MJ = 239 kcal.

However, the possibility that a small amount of n-6 fatty acids is also needed by rainbow trout cannot be discounted. Thus, Yu et al.[37] fed a semipurified diet containing 10 g linolenate/kg as the only source of EFA to rainbow trout for 34 months. During this time the fish matured, the eggs produced were hatched and the second generation of fry grown on for 3 months. The fry grew normally, no tissue pathologies were evident, and the findings were seen as supporting the thesis that n-3 series fatty acids are the only EFA needed by trout. When analyzed, the eggs and carcass phospholipids were found to contain a small quantity (0.5% by weight) of 20:4n-6. This was thought to have arisen because of incomplete extraction of diet components such as dextrin, casein, and gelatin. Nevertheless, it leaves open the possibility that n-6 fatty acids are a significant factor in the well-being of the fish.

In the same context it may be noted that, while the major phospholipids of rainbow trout (phosphatidylcholine and phosphatidylethanolamine) are rich in n-3 polyunsaturated fatty acids (PUFA), one of the minor phospholipids, i.e., phosphatidylinositol (PI), contains large amounts of 20:4n-6.[38] PI from trout tissues thus resembles that from mammals, where the fatty acid composition is closely controlled. In mammals PI is very active metabolically, being involved in the transduction of hormone signals through biomembranes. Similarly in fish, the salt-transporting epithelia in the gills are affected by the metabolic activity of PI.

Finally, there is now considerable evidence that prostaglandins derived from 20:4n-6 are active in fish, where they perform similar functions as in mammals.[39] This adds weight to the view that a small but significant requirement for n-6 fatty acids may exist in rainbow trout.

VITAMINS

The values given in Table 8 are the best current estimates of requirements. For certain vitamins, such as ascorbic acid, much larger amounts than those shown in the table have frequently been added to practical diets. This is because ascorbic acid is so susceptible to loss by oxidation. With the advent of stable forms of ascorbic acid, with high bioavailability such as the ascorbic acid phosphate, the use of large safety margins in practical diets may be obviated.

Some of the early measurements of vitamin requirements of fish indicated substantially higher requirements than those of mammals. More recent measurements on rapidly growing trout fry have given values in line with those of mammals.

MINERALS

Measurement of mineral requirements of fish is complicated by the facility with which fish can obtain some minerals from the water. The extreme case is provided by marine fish, which appear to require few dietary minerals other than phosphorus, iron, and zinc. By the same token, requirements of rainbow trout may vary with the mineral content of the water. Some values are shown in Table 8.

PRACTICAL AND SPECIAL DIETS

Most manufacturers now provide a series of diets for rainbow trout production. These essentially meet the needs of starter (fry), grower, and broodstock fish. In formulating these diets it is first necessary to decide on the desired energy density and on the protein/energy ratio. Thereafter a fixed or basal part of the feed can be formulated from nutritional information, for example, a certain minimal amount of fish meal may be necessary in trout diets, together with soybean meal and other ingredients, to balance essential amino acids. Fish oil is necessary to supply EFA, and fixed levels of vitamin and mineral premixes (formulated separately) are included. The energy and protein content of this basal component are calculated, and thereafter the required composition of the remainder of the feed formulation is inferred. The ingredients for this portion of the feed are designated, paying due attention to the overall balance of available nutrients. A full account of feed formulation, feedstuff specification, and so on is provided in the handbook by Cho et al.[38] The compositions of typical examples of trout starter, grower, and broodstock diets are shown in Table 10.

Diets are normally produced as dry pellets (less than 10% moisture) manufactured by steam pelleting. Extruded pellets are not usually necessary for trout feeding. The main factors influencing intake of organoleptically acceptable feed by trout are water temperature and energy content of the diet. Water temperature affects metabolic rate and energy expenditure; energy content of the diet affects feed consumption, because trout eat to meet an energy requirement.

Trout fry must be fed as soon as the yolk sac has been resorbed and they begin to rise for food. At this time they must be fed frequently (e.g., once per hour) during daylight hours. It is difficult to avoid overfeeding (i.e., adding excess feed to the water) at this stage and wasted feed should be removed from the water at regular intervals. The feed should be carefully screened to remove fines and over- and undersized granules and to ensure that only properly sized crumbles are offered to the fish.

As the fry grow, small pellets (2 to 3 mm) are substituted for crumbles and the frequency of feeding is reduced. Feeding guides are produced by most feed companies for different sizes

TABLE 10

Practical Diet Formulae (%) for Salmonids by University of Guelph and Ontario Ministry of Natural Resources (OMNR)[41]

Ingredient	MNR-89S[a]	MNR-89G[b]	MNR-89B[c]	C203[d]
Fish meal (68% CP, 10% fat)	40	20	30	30
Alfalfa meal (17% CP)	—	—	10	—
Blood meal (80% CP)	10	9	9	—
Brewer's yeast (45% CP)	5	—	—	—
Corn gluten meal (60% CP)	12	17	13	17
Soybean meal (49% CP)	8	12	12	13
Wheat middlings (17% CP)	—	20	8.5	16.5
Whey (12% CP)	9.5	7	7	10
Vitamin premix[e]	1.5	1	1.5	1
Mineral premix[f]	1	1	1	1
Fish oil, herring or capelin	13	13	8	11.5
Composition				
Digestible energy (DE), MJ/kg	20	17	17	17
Digestible protein:DE, g/MJ	24	22	25	22
Digestible lipid, g/kg	175	165	120	160
Total phosphorus, g/kg	9	7	8	8
Dry matter dig. coeff., %	80	70	71	71
Protein dig. coeff., %	93	94	93	93
Energy dig. coeff., %	90	82	81	83

[a] MNR-89S = high nutrient-dense/low waste diet for fry or salmonids.
[b] MNR-89G = economical grower diet for fingerling and yearling.
[c] MNR-89B = broodstock diet having higher ratio of protein to energy.
[d] C203 = reference diet for experimental research.
[e] See Table 11 for composition.
[f] See Table 12 for composition.

of fish. These may owe something to the meat meal diets of the past and may not be really applicable to the diets currently available. Caution should therefore be exercised in applying them, and it may indeed be preferable to feed juveniles and larger fish to satiation, or even slightly below, three or four times each day than to stick rigidly to a feeding guide.

During the summer or hot season, when water temperatures are high, the feeding time, feeding frequency, and amount of food given at each feed should be adjusted in accord with the concentration of dissolved oxygen in the water. Giving less food at each feeding while increasing the frequency of feedings will help prevent problems that might arise from low levels of dissolved oxygen. In addition, uneaten food may cause gill damage as well as supporting bacterial and fungal growth all of which may lead to disease.

At low water temperatures frequent feeding will not be necessary; there is a tendency when feeding only once or twice per day to feed to satiety (thereby compensating for the reduction in number of feeds). Particular attention is necessary to avoid feed waste under these circumstances. At very low water temperatures, no feed for several days will not harm the fish. In fact, the energy cost of utilizing ingested food may even tend to debilitate the fish.

Mechanical methods of delivering feed are now largely replacing hand feeding. These include (1) mechanical feeders activated by a timer and (2) pendulum or other types of demand feeders which are activated by the fish. Mechanical feeders must be calibrated and adjusted regularly as the amount of feed dispensed depends on the type and size of feed. A misconception about the demand feeder is that it never overfeeds, but this is not necessarily the case. Whether mechanical or demand feeders are being used, data on feed intake and growth should be monitored continuously to ensure that food conversion rates are in the expected range and

TABLE 11
Vitamin Premix for MNR Salmonid Diets[41]

Ingredient[a]	g/kg premix
Vitamin A	250,000 IU
Vitamin D_3	240,000 IU
Vitamin E	3,000 IU
Vitamin K	1
Vitamin B_{12}	0.002
Ascorbic acid	40
Biotin	0.02
Folic acid	0.5
Niacin	15
Pantothenic acid	4
Pyridoxine	1
Riboflavin	1
Thiamine	1
Choline chloride	100
Wheat middlings	Remainder

[a] All ingredients and premixes must be ground to pass a 0.25-mm screen.

TABLE 12
Mineral Premix for MNR Salmonid Diets[41]

Ingredient[a]	g/kg premix
Copper sulfate (25% Cu)	2.5
Ferrous sulfate (21% Fe)	6.3
Manganese sulfate (25% Mn)	8.6
Potassium iodide (76% I)	0.8
Zinc sulfate (36% Zn)	14.4
Salt (99% NaCl)	300.0
Wheat middlings	Remainder

[a] All ingredients and premixes must be ground to pass a 0.25-mm screen.

that food is not being wasted. For large production systems, a feeder that can broadcast over a large area, such as a pneumatic feeder, is desirable. Overall, however, mechanical feeders only replace manual feeding, never feeding strategy.

REFERENCES

1. **Elliot, J. M.**, The effects of temperature and ration size on the growth and energetics of salmonids in captivity, *Comp. Biochem. Physiol.*, 73B, 81, 1982.
2. **Kaushik, S. J.**, Status of European aquaculture and fish nutrition, in *The Current Status of Fish Nutrition in Aquaculture*, Takeda, M. and Watanabe, T., Eds., Tokyo University of Fisheries, Tokyo, 1990, 3.
3. **Kossmann, H.**, Present status and problems of aquaculture in the Nordic countries with special reference to fish feed, in *The Current Status of Fish Nutrition in Aquaculture*, Takeda, M. and Watanabe, T., Eds., Tokyo University of Fisheries, Tokyo, 1990, 27.

4. **Wilson, R. P.**, Aquaculture in the United States, in *The Current Status of Fish Nutrition in Aquaculture*, Takeda, M. and Watanabe, T., Eds., Tokyo University of Fisheries, Tokyo, 1990, 57.

5. **Cho, C. Y., Castledine, A. J., and Lall, S. P.**, The status of Canadian aquaculture with emphasis on formulation, quality and production of fish feeds, in *The Current Status of Fish Nutrition in Aquaculture*, Takeda, M. and Watanabe, T., Eds., Tokyo University of Fisheries, Tokyo, 1990, 67.

6. **Watanabe, T., Davy, F. B., and Nose, T.**, Aquaculture in Japan, in *The Current Status of Fish Nutrition in Aquaculture*, Takeda, M. and Watanabe, T., Eds., Tokyo University of Fisheries, Tokyo, 1990, 115.

7. **Edwards, D. J.**, *Salmon and Trout Farming in Norway*, Fishing News Books, Farnham, England, 1978.

8. **Purdom, C. E.**, Genetic techniques for control of sexuality in fish farming, *Fish. Physiol. Biochem.*, 2, 1, 1986.

9. **Thorgaard, G. H.**, Heteromorphic sex chromosomes in male rainbow trout, *Science*, 196, 900, 1977.

10. **Bye, V. J. and Lincoln, R. R.**, Commercial methods for the control of sexual maturation in rainbow trout (*Salmo gairdneri* R.), *Aquaculture*, 57, 299, 1986.

11. **Thorgaard, G. H. and Gall, G. A. E.**, Adult triploids in a rainbow trout family, *Genetics*, 93, 961, 1979.

12. **Chourrout, D.**, Thermal induction of diploid gynogenesis and triploidy in the eggs of rainbow trout (*Salmo gairdneri* Richardson), *Reprod. Nutr. Dev.*, 20, 727, 1980.

13. **Cho, C. Y.**, Feed nutrition, feeds and feeding: with special emphasis on salmonid aquaculture, *Food Rev. Int.*, 6, 333, 1990.

14. **Cowey, C. B. and Walton, M. J.**, Studies on the uptake of (^{14}C)amino acids derived from both dietary (^{14}C)protein and dietary (^{14}C)amino acids by rainbow trout, *Salmo gairdneri* Richardson, *J. Fish. Biol.*, 33, 293, 1988.

15. **Halver, J. E., Bates, L. S., and Mertz, E. T.**, Protein requirements of sockeye salmon and rainbow trout, *Fed. Proc.*, 23 (Abstr.), 1778, 1964.

16. **Zeitoun, I. H., Halver, J. E., Ullrey, D. E., and Tack, P. I.**, Influence of salinity on protein requirements of rainbow trout (*Salmo gairdneri*) fingerlings, *J. Fish. Res. Bd. Can.*, 31, 1145, 1973.

17. **Satia, B. P.**, Quantitative protein requirements of rainbow trout, *Prog. Fish-Cult.*, 36, 80, 1974.

18. **Kaushik, S. J.**, Application of a biochemical method for the estimation of amino acid needs in fish. Quantitative arginine requirements of rainbow trout in different salinities, in *Finfish Nutrition and Fishfeed Technology*, Tiews, K. and Halver, J. E., Eds., Heeneman, Berlin, 1979, 197.

19. **Walton, M. J., Cowey, C. B., Coloso, R. M., and Adron, J. M.**, Dietary requirements of rainbow trout for tryptophan, lysine and arginine determined by growth and biochemical measurements, *Fish. Biochem. Physiol.*, 2, 161, 1986.

20. **Kim, K. I., Kayes, T. B., and Amundson, C. H.**, Protein and arginine requirements of rainbow trout, *Fed. Proc.*, 42 (Abstr.), 2198, 1983.

21. **Cho, C. Y., Kaushik, S. J., and Woodward, B.**, Dietary arginine requirement for rainbow trout, *FASEB J.*, 3 (Abstr.), 2459, 1989.

22. **Ketola, H. G.**, Requirement for dietary lysine and arginine by fry of rainbow trout, *J. Animal Sci.*, 56, 101, 1983.

23. **Choo, P. S.**, Leucine Nutrition of the Rainbow Trout, M.Sc. thesis, University of Guelph, Guelph, 1990.

24. **Kim, K. I. and Kayes, T. B.**, Test diet development and lysine requirement of rainbow trout, *Fed. Proc.*, 41 (Abstr.), 716, 1982.

25. **Walton, M. J., Cowey, C. B., and Adron, J. W.**, The effect of dietary lysine levels on growth and metabolism of rainbow trout (*Salmo gairdneri*), *Br. J. Nutr.*, 52, 115, 1984.

26. **Walton, M. J., Cowey, C. B., and Adron, J. W.**, Methionine metabolism in rainbow trout fed diets of differing methionine and cystine content, *J. Nutr.*, 112, 1525, 1982.

27. **Rumsey, G. L., Page, J. W., and Scott, M. L.**, Methionine and cystine requirements of rainbow trout, *Prog. Fish-Cult.*, 45, 139, 1983.

28. **Kim, K. I., Kayes, T. B., and Amundson, C. H.**, Requirements for sulfur containing amino acids and utilization of D-methionine by rainbow trout, *Fed. Proc.*, 43 (Abstr.), 3338 1984.

29. **Walton, M. J., Coloso, R. M., Cowey, C. B., Adron, J. W., and Knox, D.**, The effects of dietary tryptophan levels on growth and metabolism of rainbow trout (*Salmo gairdneri*), *Br. J. Nutr.*, 51, 279, 1984.

30. **Kim, K. I., Kayes, T. B., and Amundson, C. H.**, Effects of dietary tryptophan levels on growth, feed/gain, carcass composition and liver glutamate dehydrogenase activity in rainbow trout (*Salmo gairdneri*), *Comp. Biochem. Physiol.*, 88B, 737, 1987.

31. **Poston, H. A. and Rumsey, G. L.**, Factors affecting dietary requirement and deficiency signs of L-tryptophan in rainbow trout, *J. Nutr.*, 113, 2568, 1983.

32. **National Research Council**, *Nutrient Requirements of Fish*, National Academy Press, Washington, D.C., 1991, in press.

33. **Ogino, C.**, Requirements of carp and rainbow trout for essential amino acids, *Bull. Jpn. Soc. Sci. Fish.*, 46, 171, 1980.

34. **Cho, C. Y., Slinger, S. J., and Bayley, H. S.**, Bioenergetics of salmonid fishes: energy intake, expenditure and productivity, *Comp. Biochem. Physiol.*, 73B, 25, 1982.
35. **Watanabe, T., Ogino, C., Koshiishi, Y., and Matsunaga, T.**, Requirement of rainbow trout for essential fatty acids, *Bull. Jpn. Soc. Sci. Fish.*, 40, 493, 1974.
36. **Takeuchi, T. and Watanabe, T.**, Dietary levels of methyl laurate and essential fatty acid requirement of rainbow trout, *Bull. Jpn. Soc. Sci. Fish.*, 43, 893, 1977.
37. **Yu, T. C., Sinnhuber, R. O., and Hendricks, J. D.**, Reproduction and survival of rainbow trout (*Salmo gairdneri*) fed linolenic acid as the only source of essential fatty acids, *Lipids*, 14, 572, 1979.
38. **Hazel, J. R.**, Influence of thermal acclimation on membrane lipid composition of rainbow trout liver, *Am. J. Physiol.*, 236, R91, 1979.
39. **Cowey, C. B.**, The nutrition of fish: the developing scene, *Nutr. Res. Rev.*, 1, 255, 1988.
40. **Cho, C. Y., Cowey, C. B., and Watanabe, T.**, *Finfish Nutrition in Asia. Methodological Approaches to Research and Development*, International Development Research Centre, Ottawa, Publication No. IDRC-233e, 1985.
41. **Cho, C. Y.**, Formulae and Specifications for Fish Feed, Fisheries Branch, Ontario Ministry of Natural Resources, Toronto, Ont., 1989.

RED DRUM, *SCIAENOPS OCELLATUS*

Edwin H. Robinson

INTRODUCTION

Red drum, commonly referred to as redfish or channel bass, is a popular sport and food fish that ranges from the Gulf of Mexico to Massachusetts on the Atlantic coast.[1] Although red drum are found along the Atlantic sea coast, the commercial fishery is centered in the Gulf of Mexico. Popularity of red drum as a food fish has increased in recent years, mainly becaused of the favorable publicity associated with blackened redfish, a Cajun recipe that is held in high esteem by culinary experts. As a result of the increased demand for red drum, a purse-seining industry began fishing large schools of red drum in the Gulf of Mexico, which quickly threatened fish stocks. In an effort to avoid depletion of red drum stocks, several states have restricted or banned commercial harvest of red drum. These actions have resulted in a shortage of red drum to meet market demands, and thus provides an opportunity for the development of red drum aquaculture.

Aquaculture of red drum is in its infancy; thus, there are few established production methods. There are numerous culturists attempting to rear red drum commercially, but most ventures have not yet been profitable, and there is not a regular supply of cultured, food-size red drum at the present time. Most of the efforts with red drum culture thus far have occurred in Texas, South Carolina, Louisiana, Florida, and Mississippi. It is too early to accurately predict the future of red drum aquaculture, but the fish has certain characteristics that make it amenable to culture. These include the ability to control its reproductive phase in captivity, adapts to prepared feeds, rapid growth, high-quality flesh, and good market appeal. Some of the disadvantages include its intolerance to low temperature, its cannibalistic nature, and its requirement for salt water for development of early life stages. A discussion of the various techniques used to produce and culture red drum can be found elsewhere.[2-7]

Production data on cultured red drum is fragmented. It is difficult to report production figures with accuracy. Very few (if any) crops of red drum have been harvested and marketed commercially. Based on limited information, it appears that red drum can be reared in earthen ponds at densities of 3000 to 5000 kg/ha. Production at this density requires that the fish be fed a complete feed containing a relatively high level of good quality protein. Production in indoor recirculating systems is not well documented and is problematic, as with any species that is reared intensively in such systems.

LABORATORY CULTURE

Larval red drum culture in the laboratory is laborious, particularly since the larvae require natural food (zooplankton) for good survival and growth.[8-10] The larvae should be maintained at a water temperature of 25 to 30°C and 25 to 30‰ salinity. They can survive at lower temperatures and salinities but growth is slowed, at least during the first 4 to 6 weeks. Larval red drum can be stocked at a density of 10 to 20 larvae/l during the first couple of weeks. Once the larvae have been converted to brine shrimp nauplii, the density should be reduced to 1 or 2 larvae/l for 2 weeks, and then further reduced to 1 larvae/2 l for the last 2 weeks of larval culture.[11] It takes approximately 4 to 6 weeks for a 2.5 mm larvae to grow to a 25 mm juvenile.

The following feeding protocol has been successfully used to rear red drum larvae.[11] At first feeding (usually about 3 d after hatching), 5 rotifers/ml are added, regardless of larval density. The number of uneaten rotifers are estimated to maintain proper food density. Larvae should be fed 2 or 3 times a day at a rate of 3 to 5 rotifers/ml for about 10 d. At this time the larvae are changed to brine shrimp nauplii at a density of 1/ml for 1 or 2 d, then the density should be reduced by 50%. On day 15, the diet should include shrimp puree mixed with a dry fry feed, as well as the brine shrimp. Continue this regime for about 5 d and then withdraw the brine shrimp over a period of 2 to 3 weeks. The fish can now be gradually weaned from the shrimp puree-feed mixture to dry feed.

Larval feeding appears straightforward, but it is a major constraint to the development of commercial culture. Because of the need for maintaining stocks of live food, considerable skill and experience are required to rear larval red drum. Also, when the larvae are switched to a dry diet, some of the fish do not adapt and mortalities can be high. Red drum do not grow as well on dry diets as on live food, at least during early life stages.

Fingerling red drum are not as troublesome to rear in the laboratory as the larvae. Most of the nutrient work with red drum has been conducted starting with 4 to 5 g fingerlings and growing them until they reach a size of 30 to 50 g. The fish have generally been reared in brackish (5 to 6‰) water-recirculating systems. Although they can survive in fresh water, survival and growth appear to be better if some salt is included. The optimum temperature is presumed to be 25°C at 30‰ salinity.[12] Nutritional studies have been conducted in brackish water with temperatures ranging from about 22 to 33°C without observable detrimental effects.

Fingerlings (1 to 2 g) that are brought into the laboratory from ponds generally have to be acclimated to prepared feeds. Initially, the fish should be maintained on live or frozen brine shrimp. After 1 or 2 d, when the fish are actively feeding on brine shrimp, a mixture of brine shrimp and salmonid starter feed should be used for 2 or 3 d.[13] Once the fish are actively feeding on the prepared feed, the brine shrimp can be discontinued. The salmonid starter feed should be continued until the fish reach an appropriate size for experimentation.

Several types of experimental diets have been fed to fingerling red drum, including practical diets, all plant protein diets, semipurified diets, and purified diets. Diets in which the primary protein source was of plant origin (soybean meal or peanut meal) and devoid in animal protein were poorly consumed.[13] When fish meal or shrimp meal was included in soybean meal diets, feed consumption improved. Red drum consume certain semipurified diets well but do not consume purified diets. For example, they readily consume a casein-based diet if a small amount of fish meal or shrimp meal is included, but consumption is poor on a casein gelatin-based diet devoid of other animal protein, even if fish oil is used as an attractant. Also, fingerling red drum do not consume amino acid test diets. Adjustment of the diet pH to 7.0 did not improve consumption.

Although red drum do not adapt well to traditional diets used for experimentation with other warmwater fish, successful experimental diets have been developed (Table 1). It appears that some animal protein is needed in red drum feeds. For example, fish meal or shrimp meal improve the palatability of diets to red drum. Fish meal prepared from red drum muscle or muscle from certain other fish appear to be particularly suited for use in experimental diets for red drum. Since there is little information concerning the micronutrient requirements of red drum, vitamin and mineral premixes suitable for other fish are generally used.

NUTRIENT REQUIREMENTS (TABLE 2)

PROTEIN AND AMINO ACIDS

Few studies have been conducted to determine the protein requirement of red drum. Lin and Arnold[18] fed a series of semimoist diets, which ranged in protein from 30 to 50%, to juveniles reared in salt water. Weight gain and feed conversion of fish fed diets containing 30 or 35% protein were depressed compared to fish fed diets containing higher levels of protein. In

TABLE 1
Examples of Experimental Diets for Red Drum[a]

	Reference			
Ingredient	14	15	16	17
Red drum muscle[b]	40.4	—	—	—
Menhaden fish meal	—	—	28.2	—
Shrimp-head meal	10	5	—	10
Egg albumin	—	20.25	—	—
Menhaden fish meal, deboned	—	25	—	—
Casein, vitamin free	—	—	13.7	38
Dextrin	19.6	18.62	26.85	26
Cellulose	10	16.13	9.25	10
Carboxymethyl cellulose	2	—	3	2
Vitamin premix[c]	3	3	3	3
Mineral premix[d]	4	4	4	4
Menhaden fish oil	10	8	12	7
Dibasic calcium phosphate	1	1	1	1

[a] Expressed as percent dry weight.
[b] Prepared by lypholizing muscle from adult fish.
[c] See Table 3.
[d] See Table 4.

TABLE 2
Summary of Nutrient Requirements for Red Drum

Nutrient	Requirement	Unit	Ref.
Energy	3.8	kcal/kg diet	16
Protein	50	% of diet	18
	35–45	% of diet	16
	40	% of diet	14
Lysine	5.3–6.6	% of protein	19
Methionine + cystine	3.3	% of protein	20
Lipid	4–7	% of diet	17
Ascorbic acid	60–75	mg/kg diet	21
Phosphorus	0.86	% of diet	15
Zinc	20–25	mg/kg diet	22

addition, survival of fish on the lower protein diets was only 50% compared to 100% survival of fish fed diets containing an excess of 35% protein. In a second study, these workers fed diets containing either 40 or 50% protein, and concluded that 50% dietary protein was optimum.

Daniels and Robinson[16] conducted two experiments to investigate the protein and energy requirements of 4 to 5 g red drum reared in brackish (6‰) water. Menhaden fish meal and casein were the protein sources used in the experimental diets. Weight gain, feed conversion, and body composition data from the first experiment indicated that 35% dietary protein was adequate for red drum reared at a water temperature ranging from 22 to 26°C. The results of the second experiment, which was conducted at a water temperature ranging from 25 to 33°C, indicated that 45% dietary protein was required for maximal growth. Recent research[14] indicates that 40% dietary protein is optimum for juveniles reared in brackish water (6‰) at a water temperature of 21 to 25°C. The fish were fed diets in which the protein was provided from red drum muscle.

The protein requirement for juvenile red drum is similar to that reported for other fish.[24,25]

There are few data on the protein requirement of red drum other than for juveniles; however, the protein requirement will likely vary with fish size and age, as well as with other factors such as water temperature and salinity.

The lysine requirement of juvenile red drum has been investigated.[19] Practical diets, which were based on peanut meal and shrimp meal proteins supplemented with crystalline L-amino acids, were fed to red drum reared in water of 6‰ salinity. Based on weight gain, feed conversion, and serum-free lysine concentrations, the lysine requirement was found to be between 5.3 and 6.6% of the dietary protein. A more precise requirement could not be ascertained from their data. The lysine requirement of other warmwater fish is between 4.4 to 5.8% of the protein; thus the suggested lysine requirement for red drum is near that reported for other fish.[25] The total sulfur amino acid requirement and cystine replacement value have been determined for juveniles reared in brackish water (6‰) and fed semipurified diets.[20] Expressed as a percentage of dietary protein, the total sulfur amino acid requirement was 3.30%. Cystine was shown to replace about 40% of methionine to meet the total sulfur amino acid requirement. Knowledge of the lysine and sulfur amino acid requirements should allow red drum feeds to be formulated on an amino acid basis. These amino acids are limiting in plant proteins, and if the requirements for lysine and sulfur amino acids are met, it can generally be assumed that other indispensable amino acids will be adequate, using feed ingredients commonly used in fish feeds.

ENERGY

Daniels and Robinson[16] studied the protein and energy requirements of juvenile red drum reared in 6‰ salinity water. Two experiments were conducted in which the experimental diets varied in protein and energy. The energy levels used in the study were 3.4, 3.8, and 4.7 kcal/g of diet. Digestible energy was estimated by subtracting the energy contributed by dietary fiber (which was considered to be indigestible) from the gross energy of the diet. Their data indicated that about 3.8 kcal/g of diet was adequate for maximum weight gain and for a desirable carcass composition, i.e., high protein and low fat. Expressed as kcal of energy/g of dietary protein, the optimum energy level was about 9.5. This is similar to the amount of digestible energy recommended for commercial catfish feeds.[23]

LIPIDS

A properly balanced feed for red drum should include lipids (fats or oils), because they are a highly digestible source of energy for warmwater fish and they provide essential fatty acids. Juvenile red drum have been reared in brackish (6‰) water and fed diets containing graded levels of lipid ranging from 1.7 to 18.7%.[17] All diets contained approximately 1.0% lipid from shrimp-head meal, with the remainder of the lipid being supplied by supplemental menhaden oil. Weight gain was best at dietary lipid levels of 7.4 and 11.2%. Fish fed diets containing 15% or greater lipid exhibited reduced weight gains and higher feed conversion ratios than fish fed diets containing lower levels of supplemental menhaden oil. Whole body lipid concentrations increased as dietary lipid increased up to 7.4%.

In a recent study,[14] juveniles fed isocaloric diets containing 10% lipid grew better than fish fed 3% lipid plus elevated levels of dextrin. This response may not be a direct effect of dietary lipid, but rather a more efficient use of fat for energy than carbohydrate.

There is little evidence to indicate which lipid source is best for use in red drum feeds. Both menhaden and soybean oils have been used up to a level of 12% without detrimental effects.[16] A mixture of oils from both plant and animal sources will most likely suffice.

CARBOHYDRATE

There have been few studies that concentrated on the level of dietary carbohydrate that can be included in red drum feeds. Experimental diets containing from 2 to greater than 35%

TABLE 3
Vitamin Premix for Use in Experimental Diets
for Red Drum[14]

Vitamin	Amount	Units
Ascorbic acid	50	g/kg
dl-calcium pantothenate	5	g/kg
Choline bitartrate	100	g/kg
Inositol	5	g/kg
Menadione	2	g/kg
Niacin	5	g/kg
Pyridoxine·HCl	1	g/kg
Riboflavin	3	g/kg
Thiamine mononitrate	0.5	g/kg
dl-α-tocopheryl acetate	2,000	IU/kg
Vitamin A acetate	100,000	IU/kg
Biotin	0.05	g/kg
Cholecalciferol	80,000	IU/kg
Folic acid	0.18	g/kg
Vitamin B$_{12}$	0.002	g/kg
Cellulose[a]	Include	g/kg

[a] Make to 1000 g.

carbohydrate have been fed in experiments designed to study other nutrient requirements without observed adverse effects, except for fatty deposits in the liver of fish fed high carbohydrate diets.[16,17] More recently, growth depression has been observed in juveniles fed diets containing 3% lipid when the remainder of the nonprotein energy was supplied from dextrin as compared to fish fed a 10% lipid diet containing low levels of dextrin. It is unclear as to which level of carbohydrate to include in red drum feeds, but the amount included in a practical feed is dictated more by available feed ingredients and feed manufacture constraints than by nutritional considerations.

VITAMINS AND MINERALS

Ascorbic acid is the only vitamin that has been studied in red drum.[21] Juveniles were fed diets containing graded levels of ascorbic acid (in the form of ascorby-2-polyphosphate). The fish were reared in brackish (5 to 6‰) water. Based on weight gain, feed conversion, liver ascorbate levels, and histological evaluation of vertebrae, the vitamin C requirement was estimated to be 60 to 75 mg/kg diet. Fish fed vitamin C-deficient diets exhibited typical signs of scurvy.

The dietary phosphorus and zinc requirements have been determined for juvenile red drum. Purified diets containing graded levels of inorganic phosphorus ranging from 0.26 to 1.31% were used to determine the phosphorus requirement. There were no significant differences in weight gain or feed conversion after 11 weeks, but phosphorus supplements up to 0.86% increased bone calcium, as well as scale ash, calcium, and phosphorus levels. Bone ash and phosphorus were not significantly affected above 0.71% dietary phosphorus. Approximately 0.86% dietary phosphorus appears to be needed for maximum tissue mineralization in juvenile red drum.[15]

The zinc requirement for juveniles reared in brackish water (6‰) and fed graded levels of zinc in purified diets was found to be 20 to 25 mg zinc/kg diet.[22] The requirement was based on weight gain, feed conversion, and tissue mineralization.

The vitamin and mineral premixes presented in Tables 3 and 4, respectively, have been used successfully for experimental work with red drum. They are based on those used in experimental diets for catfish.

TABLE 4

Mineral Premix for Use in Experimental
Diets for Red Drum[14]

Mineral	g/kg
$Ca(H_2PO_4)_2 \cdot H_2O$	136
$C_6H_{10}CaO_6 \cdot 5\ H_2O$	348.49
$FeSO_4 \cdot 7\ H_2O$	5
$MgSO_4 \cdot 7\ H_2O$	132
K_2HPO_4	240
$NaH_2PO_4 \cdot H_2O$	88
NaCl	45
$AlCl_3 \cdot 6\ H_2O$	0.15
KI	0.15
$CuSO_4 \cdot 5\ H_2O$	0.5
$MnSO_4 \cdot H_2O$	0.7
$CoCl_2 \cdot 6\ H_2O$	1
Na_2SeO_3	0.011
$ZnSO_4 \cdot 7\ H_2O$	3

TABLE 5

Example of a 36% Crude Protein Feed for
Growout of Red Drum from Approximately
150 mm to Harvest[a]

Ingredient	Percent
Soybean meal (48%)	50
Fish meal, menhaden	12
Shrimp-head meal	5
Grains or grain byproducts[b]	28
Vitamin premix[c]	Include
Mineral premix[c]	Include
Fat[d]	1.5
Dicalcium phosphate	2.4

[a] Based on available nutritional data for red drum; formulation for floating or sinking feed.
[b] Corn, milo, wheat, or a mixture of grains can be used.
[c] Meet recommendations for either catfish or salmonids.
[d] Fish oil or mixture of fish and vegetable oils may be used; spray on finished pellets.

PRACTICAL FEEDS

Presently, practical feeds can be formulated specifically for red drum based upon feeding habits, known nutritional requirements, and on nutritional data derived from studies with other fish. It appears that feed ingredients commonly used in commercial warmwater fish feeds are satisfactory for use in red drum feeds. As expected, 12 to 15% animal protein appears to be needed in red drum feeds to provide essential nutrients and to improve palatability.[13] Red drum do not readily consume feeds prepared entirely from plant protein sources. Additionally, shrimp meal appears to be highly palatable to red drum and, although feeds containing menhaden meal may be sufficient, it is recommended that 5% shrimp meal be added, if available.

It is assumed that vitamin and mineral levels recommended for commercial catfish feeds[23] or those used for salmonids are satisfactory for red drum until their vitamin and mineral requirements are identified. An example of a practical feed that has been used for grow out of red drum is given in Table 5. The feed is based upon a limited amount of nutritional data derived specifically from studies with red drum; thus, it is based largely on the requirements of other warmwater fish. Refinements will be necessary as new information becomes available.

REFERENCES

1. **Matlock, G. C.**, The life history of red drum, in *Manual on Red Drum Aquaculture*, Chamberlain, G. W., Miget, R. J., and Haby, M. G., Eds., Sea Grant College Program, Texas A&M University, College Station, 1987, I-1.
2. **Chamberlain, G. W. and McCarty, G.**, Why choose redfish, *Aquaculture Mag.*, 11, 35, 1985.
3. **Roberts, D. E., Jr.**, Photoperiod/temperature control in the commercial production of red drum (*Sciaenops ocellatus*) eggs, in *Manual on Red Drum Aquaculture*, Chamberlain, G. W., Miget, R. J., and Haby, M. G., Eds., Sea Grant College Program, Texas A&M University, College Station, 1987, II-10.
4. **Colura, R. L.**, Hormone induced strip spawning of red drum, in *Manual on Red Drum Aquaculture*, Chamberlain, G. W., Miget, R. J., and Haby, M. G., Eds., Sea Grant College Program, Texas A&M University, College Station, 1987, II-7.
5. **McCarty, D. E.**, Design and operation of a photoperiod/temperature spawning system for red drum, in *Manual on Red Drum Aquaculture*, Chamberlain, G. W., Miget, R. J., and Haby, M. G., Eds., Sea Grant College Program, Texas A&M University, College Station, 1987, II-27.
6. **Henderson-Arzapalo, A.**, Red drum egg and larval incubation, in *Manual on Red Drum Aquaculture*, Chamberlain, G. W., Miget, R. J., and Haby, M. G., Eds., Sea Grant College Program, Texas A&M University, College Station, 1987, II-40.
7. **Holt, G. J.**, Growth and development of red drum eggs and larvae, in *Manual on Red Drum Aquaculture*, Chamberlain, G. W., Miget, R. J., and Haby, M. G., Eds., Sea Grant College Program, Texas A&M University, College Station, 1987, II-32.
8. **Treece, G. D. and Wohlschlag, N.**, Raising food organisms for intensive larval culture: algae culture, in *Manual on Red Drum Aquaculture*, Chamberlain, G. W., Miget, R. J., and Haby, M. G., Eds., Sea Grant College Program, Texas A&M University, College Station, 1987, III-6.
9. **Wohlschlag, N. S., Maotang, S., and Arnold, C. R.**, Raising food organisms for intensive larval culture: rotifers, in *Manual on Red Drum Aquaculture*, Chamberlain, G. W., Miget, R. J., and Haby, M. G., Eds., Sea Grant College Program, Texas A&M University, College Station, 1987, III-24.
10. **Treece, G. D. and Wohlschlag, N.**, Raising food organisms for intensive larval culture: artemia, in *Manual on Red Drum Aquaculture*, Chamberlain, G. W., Miget, R. J., and Haby, M. G., Eds., Sea Grant College Program, Texas A&M University, College Station, 1987, III-33.
11. **Holt, G. J., Arnold, C. R., and Riley, C. M.**, Intensive culture of larval and post-larval red drum, in *Manual on Red Drum Aquaculture*, Chamberlain, G. W., Miget, R. J., and Haby, M. G., Eds., Sea Grant College Program, Texas A&M University, College Station, 1987, III-1.
12. **Neil, W. H.**, Environmental requirements of red drum, in *Manual on Red Drum Aquaculture*, Chamberlain, G. W., Midget, R. J., and Haby, M. G., Eds., Sea Grant College Program, Texas A&M University, College Station, 1987, IV-1.
13. **Robinson, E. H., Daniels, W. H., Williams, C. D., and Warts, W. A.**, Diet development and environmental ion requirements for fingerling red fish (*Sciaenops ocellatus*), in *Proc. 1984 Fish Farm. Conf. and Ann. Conven.*, Johnson, S. K., Ed., Texas Agricultural Extension Service, College Station, 1984, 44.
14. **Serrano, J. and Gatlin, D. M., III**, Protein and energy nutrition of the red drum, *Sciaenops ocellatus*, *Aquaculture*, submitted.
15. **Davis, D. A. and Robinson, E. H.**, Dietary phosphorus requirement of juvenile red drum *Sciaenops ocellatus*, *J. World Aquaculture Soc.*, 18, 129, 1987.
16. **Daniels, W. H. and Robinson, E. H.**, Protein and energy requirements of juvenile red drum (*Sciaenops ocellatus*), *Aquaculture*, 53, 243, 1986.

17. **Williams, C. D. and Robinson, E. H.**, Response of red drum to various dietary levels of menhaden oil, *Aquaculture*, 70, 107, 1988.
18. **Lin, B. and Arnold, C. R.**, The growth response of red fish (*Sciaenops ocellatus*), Ann. World Maricul. Soc. Meeting, Washington, D.C., 1983 (Abstr.).
19. **Brown, P. B., Davis, D. A., and Robinson, E. H.**, An estimate of the dietary lysine requirement of juvenile red drum *Sciaenops ocellatus*, *J. World Aquaculture Soc.*, 19, 109, 1988.
20. **Moon, H. Y.**, Amino Acid Nutrition of the Red Drum (*Sciaenops ocellatus*): Development of an Improved Test Diet and Determination of the Total Sulfur Amino Acid Requirement, M.S. thesis, Texas A&M University, College Station, 1990.
21. **Collins, J.**, The Vitamin C Requirement of Red Drum, M.S. thesis, Mississippi State University, Mississippi State, 1990.
22. **Gatlin, D. M., III, O'Connell, and Scarpa, J.**, Dietary zinc requirement of the red drum, *Sciaenops ocellatus*, *Aquaculture*, submitted.
23. **Robinson, E. H.**, Channel catfish nutrition, *Rev. Aquatic Sci.*, 1, 365, 1989.
24. **National Research Council**, *Nutrient Requirements of Coldwater Fishes*, National Academy of Science, Washington, D.C., 1981.
25. **National Research Council**, *Nutrient Requirements of Warmwater Fishes and Shellfishes*, National Academy Press, Washington, D.C., 1983.

STURGEON, *ACIPENSER* spp.

Silas S. O. Hung

INTRODUCTION

Sturgeon are a primitive stock of teleost fish evolving approximately 250 million years ago and are a chondrostean fish in the Acipenseriformes order.[1] There are 23 species in the sturgeon family; 2 species in *Huso*, 2 species in *Scaphirhynchus*, 3 species in *Pseudoscaphirhynchus*, and 16 species in *Acipenser*.[2] Some species are marine, some go into fresh water to spawn, and some are landlocked in fresh water. Beluga sturgeon (*H. huso*) is the largest freshwater fish and a maximum body weight of 2000 kg has been reported. White sturgeon (*A. transmontanus*) is the largest among North American species, and a record weight of 816 kg has been reported. Shovelnose (*P. kaufmanni*, *P. hermanni*, and *P. fedschenkovi*) in the Aral Sea are the smallest sturgeon, with a maximum weight of 0.5 kg.[3] These fish are found only in the northern hemisphere and live mainly in temperate water, but can be found from the Arctic Circle to the subtropics.[2]

In the last 30 years many hatcheries have been built to raise sturgeon fingerlings to replenish the natural populations. In 1987, there were 26 hatcheries in the USSR and 100 to 140 million 2 to 3 g sturgeon fingerlings of Beluga (*H. huso*), Russian (*A. guldenstadti*), and Sevrjuga (*A. stellatus*) were released.[4] Hybrids of *H. huso* and *A. ruthenus* are being cultured in ponds and cages to a market size of 0.8 kg as foodfish in the USSR. Annual production of the hybrids was between 100 and 200 t in 1979.[3]

Outside the USSR, a Siberian sturgeon (*A. baeri*) farm has been established in France.[5,6] A small number of hatcheries have been established in the USA to raise lake (*A. fulvescens*), shortnose (*A. brevirostrum*), Atlantic (*A. oxyrhynchus*), and white sturgeon (*A. transmontanus*) to replenish natural stocks.[4] There are also 10 to 20 commercial scale fish farms in California and one in Northern Italy (Calvisano), which are producing white sturgeon as foodfish.[7] Viable domestic male white sturgeon have been established by the Department of Animal Science, University of California, Davis.[8]

At present, there are about 47 registered white sturgeon growers, and 11 farmers working to develop domestic broodstock in California. Several of these farmers are marketing 3.5 to 5.5 kg white sturgeon as foodfish at 2 to 3 years of age. In 1990 to 1991, a projected 500 t of white sturgeon will be produced by the California farms and 300 t by the Italian farm.[7]

LABORATORY CULTURE

In our laboratory,[9-12] sturgeon are kept in circular fiberglass tanks supplied with either surface irrigation water or aerated well water at temperatures of 8 to 30°C and 18 to 21°C, respectively. Each tank is equipped with an airstone and a screened central drain leading to an external standpipe.

Newly hatched larvae are obtained from local producers, transported to our laboratory and kept in fiberglass tanks supplied with surface irrigation water at 14 to 17°C. Ten to twelve day posthatch sturgeon are used for larval studies. The larvae are fed to satiation on live organisms or 10% of body weight/day (BW/d) of experimental diets initially. Feeding rates are reduced gradually and experiments usually last between 30 and 60 d.

To raise fingerlings for growth trials, larvae are fed a commercial semimoist trout starter feed for approximately 10 weeks. The biomass of the tanks is reduced periodically while the water flow rate is increased as the larvae grow larger. Upon reaching an average size of 10 to 15 g, 300 to 400 fingerlings are transferred to a larger tank supplied with aerated well water

TABLE 1
Example of a Purified Diet for White Sturgeon[9,10]

Ingredient	Percent
Casein	31
Wheat gluten	15
Egg white	4
Dextrin	28
Cod liver oil	4
Corn oil	4
Lard	4
Vitamin mix[a]	4
Mineral mix[b]	3
Cellulose	3

[a] To provide as mg/kg of diet: thiamine·HCl, 2000; riboflavin, 200; nicotinic acid, 1000; Ca-d-pantothenate, 1200; pyridoxine·HCl, 120; cobalamin, 0.12; folic acid, 80; d-biotin (1%), 2000; choline chloride, 6000; l-inositol, 4000; L-ascorbic acid, 1000; p-amino-benzoic acid, 1200; retinyl acetate (500,000 IU/g), 80; cholecalciferol (1%), 280; DL-α-tocopheryl acetate (250 IU/g), 3000; menadione, 100; ethoxyquin, 120; BHA, 40.

[b] To provide as mg/kg of diet: calcium carbonate, 630; calcium phosphate 22,050; citric acid, 68.1; cupric citrate·2.5H_2O, 13.8; ferric citrate·5H_2O, 167.4; magnesium oxide, 750; manganese citrate, 250.5; potassium iodide, 0.3; potassium phosphate dibasic, 2430; potassium sulfate, 2040; sodium chloride, 918; sodium phosphate, 64.2; zinc citrate·2H_2O, 39.9.

at a rate of 14 to 16 l/min and at a temperature of 18 to 21°C. The fish are gradually weaned during the next 10 days from the commercial feed to our standard purified diet (Table 1). Prior to the start of each experiment, 20 to 30 fish are transferred to smaller fiberglass tanks and acclimated for 10 to 14 d. Our feeding studies usually last 8 weeks and the fish are fed at 2% of BW/d. Fish are weighed once every 2 weeks and the amount of feed is adjusted accordingly. To minimize stress after weighing, feeding is discontinued for 24 h. The fish are also given a static bath of 10 ppm of active nitrofurazone for an hour postweighing to prevent possible bacterial infestations caused by handling.

NUTRIENT REQUIREMENTS

Very little information is available on the nutrient requirements of different species of sturgeon. The lack of information has been cited as a major limiting factor for future development of sturgeon aquaculture. Furthermore, because of the lack of a suitable commercial feed, live organisms or salmonid feeds have been used to raise sturgeon. Nutritional deficiencies have been observed after prolonged feeding of live organisms, especially when a single live food organism was used as the predominant source of food. Prolonged feeding of white sturgeon with salmonid feeds can also result in poor growth and abnormal symptoms, which may be nutritionally related.

It is difficult or, perhaps inaccurate, to extrapolate information from other species of fish to sturgeon because of their primitiveness on the evolutionary scale and their unique biological characteristics.[1-3] These unique characteristics include a collagenous bone structure, a very long life span (up to or over 100 years), a late sexual maturation (10 years for wild males and 15 years for females), and a long period between spawning (2 to 3 years in wild males and 4 to 5 years in females). Besides serving sturgeon farmers and feed manufacturers, information on the nutrient requirements of sturgeon are of interest to fish biologists and comparative nutritionists. Comparing the nutrition of sturgeon to modern teleost will provide insights into the effect of evolution on the nutrition of fish.

Some larval feeding studies have been conducted with Siberian and white sturgeon.[13-18] These studies have compared the growth performance of larvae fed different live organisms with those fed dry or semimoist feeds. It is difficult to extract information on nutrient requirements from these studies. Limited nutritional information, however, has been acquired recently for white sturgeon from growth trials under controlled conditions and with defined experimental diets.[12]

PROTEIN AND AMINO ACIDS

Minimum dietary protein requirements for maximum growth of Siberian and white sturgeon have been reported to be $40.5 \pm 1.6\%$ and $40 \pm 2\%$, respectively.[19,20] Dietary protein levels of 49 and 50%, respectively, were required for maximum growth. It is of interest that these requirement values are so similar for the two species because major differences existed between the two studies. These differences included fish size (22 vs. 145 g), protein sources (practical vs. purified ingredients), and length of the growth studies (12 vs. 8 weeks for the Siberian and white sturgeon, respectively). The only information available on the quality of dietary proteins is for white sturgeon. Protein quality was evaluated based on 8-week percent energy retained data. The results ranked the test proteins in decreasing order as casein:wheat gluten:egg white (62:30:8) = casein > defatted shrimp meal > defatted herring meal > soybean concentrate > egg white > gelatin > defatted zein meal.[11]

An essential amino acid (EAA) requirement profile has been estimated for juvenile Siberian sturgeon using daily whole-body EAA increments[19] and is presented in Table 2. Although there is no information on the EAA requirements of white sturgeon, whole-body and egg amino acid profiles have been determined (Table 3). Whole-body amino acid composition of small (3 month old, 17.3 ± 3.1 g) and large (14 month old, 999.3 ± 41.6 g) white sturgeon were similar. This was also true for small (44.7 g) and large (1000 g) Siberian sturgeon. Similar observations have been reported for channel catfish.[21] Whole-body amino acid compositions, however, were quite different between the two sturgeon species. This is in contrast with some other fish because whole-body amino acid composition of rainbow trout, Atlantic salmon, coho salmon, and cherry salmon were very similar.[22] It has been shown that whole-body EAA profiles of common carp and channel catfish closely resemble their EAA requirement profiles.[21,23] Muscle EAA profiles have also been suggested as a good indicator for the EAA requirements.[24] Therefore, whole-body and muscle EAA profiles have been used to estimate the EAA requirements of goldfish, golden shiner, fathead minnow, and dolphin fish.[25,26] Similarly, EAA profiles of fish eggs may be used to estimate the EAA requirements in larval fish.

LIPIDS

No information is available on the optimum dietary lipid level for maximum growth of sturgeon. Commercial salmonid feeds commonly used to raise sturgeon contain between 10 and 20% lipid. Due to the high level of fish oil, these feeds usually contain a high level of long-chain highly unsaturated fatty acids. Salmonid feeds with a higher lipid content are used for sturgeon larvae and broodstock.

A recent study in our laboratory showed no growth difference between white sturgeon fed 15% of one of eight different lipid sources for 8 weeks. The lipid sources were control (corn oil:cod liver oil:lard, 1:1:1), corn oil, cod liver oil, lard, soybean oil, sunflower oil, canola oil, and linseed oil. The lack of any growth difference may have resulted from a large storage of essential fatty acids (EFA) in the fingerlings (initial BW of 32 to 39 g), a low EFA requirement of white sturgeon, or the short duration of the growth trial.

TABLE 2
Whole-Body Amino Acid Composition and Estimated Essential Amino Acid Requirements of Siberian Sturgeon[19]

Amino acid[a]	Body weight (g)		Requirement as % of dietary protein
	44.7	1000	
Arginine	6.7 ± 0.3	7.7 ± 0.4	2.8 ± 0.6
Lysine	11.7 ± 1.8	11.1 ± 1.1	6.3 ± 2.0
Histidine	2.0 ± 0.1	2.4 ± 0.1	1.1 ± 0.4
Isoleucine	5.1 ± 0.3	5.4 ± 1.2	2.1 ± 0.6
Leucine	6.8 ± 0.8	7.4 ± 1.8	3.2 ± 1.8
Valine	5.3 ± 0.2	6.0 ± 1.8	2.3 ± 0.6
Phenylalanine	4.1 ± 0.1	3.8 ± 0.5	1.5 ± 0.4
Threonine	5.0 ± 0.2	5.4 ± 0.8	2.2 ± 0.6
Tyrosine	4.0 ± 0.2	5.3 ± 0.8	—
Methionine	NR[b]	NR	—
Tryptophan	NR	NR	—
Aspartic acid	8.4 ± 3.1	10.2 ± 2.1	—
Glutamic acid	18.5 ± 0.1	18.3 ± 1.6	—
Serine	4.8 ± 0.1	5.6 ± 0.7	—
Glycine	9.2 ± 0.4	11.9 ± 1.7	—
Alanine	7.6 ± 0.1	7.3 ± 2.7	—
Proline	NR	NR	—

[a] Expressed as g/16 g nitrogen.
[b] Not reported.

TABLE 3
Whole-Body and Egg Amino Acid Composition of White Sturgeon[a]

Amino acid[b]	Body weight (g)		Eggs
	17.3 ± 3.1	993.3 ± 41.6	
Arginine	6.3 ± 0.1	6.5 ± 0.1	6.8 ± 0.1
Lysine	7.9 ± 0.1	7.6 ± 0.1	8.0 ± 0.1
Histidine	2.3 ± 0.2	2.9 ± 0.1	2.7 ± 0.1
Isoleucine	2.8 ± 0.1	2.5 ± 0.1	4.2 ± 0.1
Leucine	7.0 ± 0.1	6.3 ± 0.1	8.0 ± 0.1
Valine	2.9 ± 0.1	2.7 ± 0.1	4.9 ± 0.1
Phenylalanine	3.6 ± 0.1	3.5 ± 0.1	3.7 ± 0.1
Threonine	4.3 ± 0.1	4.1 ± 0.1	5.3 ± 0.3
Tyrosine	3.3 ± 0.1	2.8 ± 0.1	4.0 ± 0.1
Methionine	3.1 ± 0.1	2.9 ± 0.1	3.1 ± 0.1
Tryptophan	1.1 ± 0.1	1.0 ± 0.1	1.2 ± 0.1
Aspartic acid	11.1 ± 0.2	10.7 ± 0.1	9.3 ± 0.1
Glutamic acid	15.2 ± 0.3	14.2 ± 0.1	13.7 ± 0.1
Serine	5.6 ± 0.3	6.0 ± 0.3	8.2 ± 0.1
Glycine	8.9 ± 0.2	10.8 ± 0.2	3.1 ± 0.1
Alanine	6.9 ± 0.2	7.5 ± 0.1	6.4 ± 0.1
Proline	6.0 ± 0.1	6.6 ± 0.1	5.4 ± 0.1

[a] Unpublished data.
[b] Expressed as g/100 g amino acids.

CARBOHYDRATES

Growth, feed gain ratio, and protein retention efficiency were used to compare the ability of Siberian sturgeon to utilize raw, gelatinized, and extruded starch.[27] The utilization of gelatinized and extruded starch was about the same but better than raw starch when fed at a dietary level of 38%.

Percent energy retained was used to compare the ability of white sturgeon to utilize diets containing 27.2% of one of 8 carbohydrates.[28] Ability of sturgeon to utilize different carhohydrates in decreasing order was: glucose = maltose > raw corn starch = dextrin = sucrose > lactose = fructose = cellulose. The poor utilization of either raw corn starch or dextrin may have been the result of low intestinal α-amylase activity. The poor utilization of fructose was due to poor fructose absorption.[29] The poor utilization of sucrose and lactose was primarily due to the low intestinal sucrase and lactase activities.[28] No abnormal hepatic or intestinal histology was observed in the fish fed the glucose or maltose diets.[30] However, fish fed diets containing sucrose, lactose, or fructose had an increased amount of lumenal water in their distal intestines. This effect was similar to that observed in the large intestine of higher vertebrates with diarrhea secondary to disaccharide intolerance.

VITAMINS AND MINERALS

The only two vitamins that have been studied in sturgeon are ascorbic acid and choline. Two approaches have been used to study the ascorbic acid requirement of white sturgeon.[31] These involved the measurement of tissue acitivities of gulonolactone oxidase, a key enzyme for *de novo* synthesis of ascorbic acid, and a growth trial with fish fed graded levels of ascorbic acid. Both studies indicate that there is no ascorbic acid requirement by white sturgeon for good growth and maintenance of tissue levels.

The choline requirement of juvenile white sturgeon was found to be 0.17 to 0.32% (30 to 60 mg/kg BW/d).[32,33] Due to various factors that may affect the choline requirement of sturgeon, a level of 0.4 to 0.6% is recommended as a safe supplemental level for sturgeon feeds. There is no requirement for lecithin by white sturgeon, but in the absence of dietary choline, 8% refined soy lecithin can support good growth. Furthermore, this level of refined soy lecithin may have some beneficial effect on the growth and survival of white sturgeon larvae.[12]

The vitamin mix in our purified test diets (Table 1) contain levels 5 to 10 times higher than those recommended for channel catfish. Supplementing excessively high levels of vitamins was an attempt to overcome the leaching problem because sturgeon feed very slowly. The leaching problem, however, has been shown to be much less than anticipated,[34] and our vitamin mix is definitely too high. It is not certain whether similar problems exist in our mineral mix. There is, however, no information on the mineral requirements of sturgeon.

PRACTICAL FEEDS AND FEEDING

Sturgeon is a relatively new species for aquaculture, and there is a paucity of information on the optimum production conditions of sturgeon. Among these, the lack of information on the nutrient requirements and optimum feeding rates are the most critical. At present, there is no suitable commercial sturgeon feed on the market and sturgeon are fed feeds designed mainly for salmonids. An example of a prototype commercial sturgeon feed formulation is presented in Table 4.

Sturgeon farmers in California prefer to use semimoist salmonid starter feeds for their larvae (8 to 14-d posthatch to 2 to 3 g) because sturgeon are known to prefer softer feeds. It is not certain whether the high fat content in the feeds is beneficial to larvae, although

TABLE 4
Formulation of a Prototype Commercial
Sturgeon Feed

Ingredient	Percent
Anchovy fish meal (66% CP)	60.0
Rice bran	15.0
Bakery dry product	10.0
Fish oil	6.4
Wheat mill	5.2
Perma pel	2.0
Choline chloride	0.9
Vitamin premix[a]	0.4
Trace mineral premix[b]	0.1
Mold inhibitor	0.1

[a] To supply the following in IU or mg/kg of diet: retinyl acetate, 6600 IU; cholecalciferol, 400,000 IU; α-tocopheryl acetate, 330 mg; niacin, 193 mg; riboflavin, 53 mg; pyridoxine·HCl, 31 mg; thiamine mononitrate, 35 mg; Ca-d-pantothenic acid, 99 mg; folic acid, 8.8 mg; biotin, 0.55 mg; vitamin B_{12}, 0.033 mg; vitamin C, 963 mg; ethoxyquin, 138 mg; MPB, 22 mg.
[b] Supplied the following in mg/kg of diet: zinc, 165; iron, 48.4; manganese, 27.5; iodine, 5.5; and copper, 5.5; all minerals were supplied in the sulfate form, except that iodine was supplied as E.D.D.I.

increased intestinal lipolytic activity has been observed in 10- to 30-d posthatch larvae.[35] Fatty liver has been observed in fingerlings in our laboratory, as well as from several farms. The fatty liver, however, does not seem to affect the growth or health of the fingerlings.

Regular dry salmonid feeds from crumbles to 6.4 mm pellets are used for fingerlings and growout. Although the size of the mouth relative to body of sturgeon is quite large, these fish prefer smaller crumbles and pellets than other fish of similar body sizes.

Salmonid broodstock feeds with high lipid content have been used to raise captive white sturgeon broodstocks. Males fed these feeds reach sexual maturity and produce viable sperm at 3 to 5 years of age with a body size of 5 kg or more. Females fed the same feeds, however, fail to reach sexual maturity up to 9 years of age. Nutrition and environmental conditions are suspected to play a major role in the failure of reproduction in these females. Recently, sexual maturation and subsequent successful spawning have been achieved with 6- to 7-year-old captive females that had been subjected to low winter temperatures (7 to 9°C) and fed krill, minced squid, minced fish, or a very high lipid salmonid broodstock feed.[8]

Various feeding techniques such as hand feeding, automatic feeders, and demand feeders have been used to raise sturgeon. Hand and automatic feeders are used mainly for larval rearing. There is no information on the optimum feeding rate of sturgeon larvae, but up to 25 to 30% of BW/d as semimoist salmonid starter feeds are commonly fed to the larvae after they have initiated exogenous feeding (about 10 d posthatch). Feeds are divided into equal portions and fed to larvae once every 2 to 3 h with time-controlled automatic feeders. Others prefer to hand feed larvae to satiation once every 2 h. Some farmers even hand feed during working hours and use automatic feeders during the rest of the day. Excess feed in the tanks is removed once or twice per day. Feeding rates are decreased gradually to 3 to 4% in the next 30 d.

Demand feeders are the most common technique for raising fingerlings and growouts. Feeds are usually divided into two portions and feeders filled in early morning and late afternoon.

TABLE 5
Recommended Feeding Rates for White Sturgeon

Body weight (g)	Temperature (°C)	Feeding rate (% BW/d)	Ref.
125	11	0.5	UD[a]
125	18	1.5	UD
30	20	2.0	10
30	26	2.5	UD
30	23	3.0	UD
250	18	1.5	36
760	22	1.3	UD

[a] Unpublished data.

Information on the optimum feeding rate of sturgeon is very important to sturgeon farmers because maintaining fish at the optimum feeding rate provides the greatest growth of fish per unit of feed cost. Five studies have been conducted to determine the effects of feeding rates on growth and feed efficiency of different sizes of white sturgeon at different temperatures. The results of these studies are summarized in Table 5.

ACKNOWLEDGMENTS

I wish to thank Drs. D. E. Conklin, F. S. Conte, S. I. Doroshov, S. J. Kaushik, and Mr. Ken Beer for providing some of the most recent information on sturgeon nutrition and aquaculture. Critical review during preparation of this review by Drs. Conte and Doroshov is also appreciated.

REFERENCES

1. **Sewertzoff, A. N.**, The place of cartilaginous gonoids in the system and the evolution of the *Osteichthyes*, *J. Morphol.*, 38, 105, 1923.
2. **Scott, W. B. and Crossman, E. J.**, Sturgeon family — Acipenseridae, in *Freshwater Fishes of Canada*, Bull. 184, Fisheries Research Board of Canada, Ottawa, 1973, 78.
3. **Doroshov, S. I.**, Biology and culture of sturgeon: aciperseriformes, in *Advances in Aquaculture*, Vol. 2, Muir, J. F. and Roberts, R. J., Eds., Westview Press, Boulder, CO, 1985, 251.
4. **Doroshov, S. I. and Binkowski, F. P.**, Epilogue: a perspective on sturgeon culture, in *North American Sturgeons: Biology and Aquaculture Potential*, Doroshov, S. I. and Binkowski, F. P., Eds., Dr. W. Junk Publishers, Dordrecht, Netherlands, 1985, 147.
5. **Barrucand, M., Ferlin, P., and Sabaut, J. J.**, Alimentation artificielle de l'esturgeon (*Acipenser baeri*), in *Finfish Nutrition and Finfish Technology*, Halver, J. E. and Tiews, K., Eds., Vol. I., Heenneman, Berlin, 1979, 411.
6. **Williot, P., Brun, R., Rouault, T., and Rooryck, O.**, Management of female breeders of Siberian sturgeon *Acipenser baeri* Brandt: preliminary results, in *Proc. First Int. Symp. on Sturgeon*, CEMAGREF, Bordeaux, France, 1991, 365.
7. **Beer, K. E.**, personal communication.
8. **Doroshov, S. I.**, personal communication.
9. **Hung, S. S. O., Moore, B. J., Bordner, C. E., and Conte, F. S.**, Growth of juvenile white sturgeon (*Acipenser transmontanus*) fed different purified diets, *J. Nutr.*, 117, 328, 1987.
10. **Hung, S. S. O. and Lutes, P. B.**, Optimum feeding rate of hatchery-produced juvenile white sturgeon (*Acipenser transmontanus*): at 20°C, *Aquaculture*, 65, 307, 1987.
11. **Stuart, J. S. and Hung, S. S. O.**, Growth of juvenile white sturgeon (*Acipenser transmontanus*) fed different proteins, *Aquaculture*, 76, 303, 1989.

12. **Lutes, P. B., Hung, S. S. O., and Conte, F. S.**, Survival, growth, and body composition of white sturgeon larvae fed purified and commercial diets at 14.7 and 18.4°C, *Prog. Fish-Cult.*, 52, 192, 1990.
13. **Monaco, G., Buddington, R. K., and Doroshov, S. I.**, Growth of white sturgeon (*Acipenser transmontanus*) under hatchery conditions, *J. World Maricul. Soc.*, 12, 113, 1981.
14. **Buddington, R. K. and Doroshov, S. I.**, Feeding trials with hatchery produced white sturgeon juvenile (*Acipenser transmontanus*), *Aquaculture*, 36, 237, 1984.
15. **Lindbery, J. C. and Doroshov, S. I.**, Effect of diet switch between natural and prepared foods on growth and survival of white sturgeon juveniles, *Trans. Am. Fish. Soc.*, 115, 166, 1986.
16. **Dabrowski, K., Kaushik, S. J., and Fauconneau, B.**, Rearing of sturgeon (*Acipenser baeri* Brandt) larvae. I. Feeding trial, *Aquaculture*, 47, 185, 1985.
17. **Fauconneau, B., Aguirre, P., Dabrowski, K., and Kaushik, S. J.**, Rearing of sturgeon (*Acipenser baeri* Brandt) larvae. 2. Protein metabolism: influence of fasting and diet quality, *Aquaculture*, 51, 117, 1986.
18. **Dabrowski, K., Kaushik, S. J., and Fauconneau, B.**, Rearing of sturgeon (*Acipenser baeri* Brandt) larvae. III. Nitrogen and energy metabolism and amino acid absorption, *Aquaculture*, 65, 31, 1987.
19. **Kaushik, S. J., Breque, J., and Blanc, D.**, Requirements for protein and essential amino acids and their utilization by Siberian sturgeon (*Acipenser baeri*), in *Proc. First Int. Symp. on Sturgeon*, CEMAGREF, Bordeaux, France, 1991, 25.
20. **Moore, B. J., Hung, S. S. O., and Medrano, J. F.**, Protein requirement of hatchery-produced juvenile white sturgeon (*Acipenser transmontanus*), *Aquaculture*, 71, 235, 1988.
21. **Wilson, R. P. and Poe, W. E.**, Relationship of whole body and egg essential amino acid patterns to amino acid requirement patterns in channel catfish, *Ictalurus punctatus*, *Comp. Biochem. Physiol.*, 80B, 385, 1985.
22. **Wilson, R. P. and Cowey, C. B.**, Amino acid composition of whole body tissue of rainbow trout and Atlantic salmon, *Aquaculture*, 48, 373, 1985.
23. **Cowey, C. B. and Tacon, A. G. J.**, Fish nutrition — relevance to invertebrates, in *Proc. Second Int. Conf. Aquaculture Nutr.: Biochem. Physiol. Approaches to Shellfish Nutr.*, Pruder, G. D., Langon, D. J., and Conklin, D. E., Eds., Louisiana State Univ., Division of Continuing Education, Baton Rouge, LA, 1983, 13.
24. **Cowey, C. B. and Luquet, P.**, Physiological basis of protein requirements of fishes. Critical analysis of allowances, in *Protein Metabolism and Nutrition*, Pion, R., Arnal, M., and Bonin, D., Eds., Vol. 1, INRA, Paris, 1983, 365.
25. **Gatlin, D. M., III**, Whole-body amino acid composition and comparative aspects of amino acid nutrition of the goldfish, golden shiner and fathead minnow, *Aquaculture*, 60, 223, 1987.
26. **Ostrowski, A. C. and Divakaran, S.**, The amino acid and fatty acid compositions of selected tissues of the dolphin fish (*Coryphaena hippurus*) and their nutritional implications, *Aquaculture*, 80, 285, 1989.
27. **Kaushik, S. J., Luquet, P., Blanc, D., and Paba, A.**, Studies on the nutrition of Siberian sturgeon, *Acipenser baeri*. 1.Utilization of digestible carbohydrates by sturgeon, *Aquaculture*, 76, 97, 1989.
28. **Hung, S. S. O., Fynn-Aikins, F. K., Lutes, P. B., and Xu, R. P.**, Ability of juvenile white sturgeon (*Acipenser transmontanus*) to utilize different carbohydrates, *J. Nutr.*, 119, 727, 1989.
29. **Hung, S. S. O.**, Carbohydrate utilization by white sturgeon as assessed by oral administration tests, *J. Nutr.*, in press.
30. **Hung, S. S. O., Groff, J. M., Lutes, P. B., and Fynn-Aikins, F. K.**, Hepatic and intestinal histology of juvenile white sturgeon fed different carbohydrate sources, *Aquaculture*, 87, 349, 1990.
31. **Conklin, D. E.**, personal communication.
32. **Hung, S. S. O. and Lutes, P. B.**, A preliminary study on the non-essentiality of lecithin for hatchery-produced juvenile white sturgeon (*Acipenser transmontanus*), *Aquaculture*, 68, 353, 1988.
33. **Hung, S. S. O.**, Choline requirement of hatchery-produced juvenile white sturgeon (*Acipenser transmontanus*), *Aquaculture*, 78, 183, 1989.
34. **Bordner, C. E., D'Abramo, L. R., Conklin, D. E., and Baum, N. A.**, Development and evaluation of diets for crustacean aquaculture, *J. World Aquacult. Soc.*, 17, 44, 1986.
35. **Buddington, R. K. and Doroshow, S. I.**, Development of digestive secretions in white sturgeon juveniles (*Acipenser transmontanus*), *Comp. Biochem. Physiol.*, 83A, 233, 1986.
36. **Hung, S. S. O., Lutes, P. B., Conte, F. S., and Storebakken, T.**, Growth and feed efficiency of white sturgeon (*Acipenser transmontanus*) sub-yearlings at different feeding rates, *Aquaculture*, 80, 147, 1989.

TEMPERATE BASSES, *MORONE* spp., AND
BLACK BASSES, *MICROPTERUS* spp.

Thomas M. Brandt

INTRODUCTION

STRIPED AND HYBRID STRIPED BASSES

The original distribution of the striped bass, *Morone saxatillis*, is believed to have extended from the Lawrence River, New Brunswick, south along the east coast of North America, around the tip of Florida, and into the Gulf of Mexico as far as Louisiana.[1] The striped bass was a marine and estuarine coastal species until stocking extended its range across much of the United States and into the Pacific Ocean.[1] Striped bass have also been stocked in the Soviet Union, France, and Portugal.[2] One of the first reports of hatchery production of hybrid striped bass (striped bass × white bass, *Morone chrysops*) and the reciprocal cross was by Bishop in 1968.[3]

A severe decline in striped bass populations along both the east and west coasts of North America during the 1970s and 1980s precipiated a doubling of production of striped bass at federal hatcheries from under 2 million in 1980 to over 5 million in 1989. Weight produced during this period increased tenfold to about 45,000 kg in 1989.[4] Striped bass have been primarily produced for restocking of public waters. Presently, most hybrid striped basses are produced for human consumption. The private production of striped and hybrid striped basses is located primarily in four southeastern states and California. Total annual production is about 0.7 million kg.[5] Laws regulating the production of striped and hybrid basses vary widely. In some states, the commercial culture of striped bass is illegal.

Methods used to culture the striped bass and its hybrids vary greatly from pond culture to indoor recirculation systems.[6-13] Wild broodstock are either spawned manually or allowed to spawn in circular tanks. Fertilized eggs are incubated in jars or tanks. First-feeding larvae are provided live brine shrimp or zooplankton. The best time to switch fry from live to artificial feed has not been determined. Generally, there is a transition period when both live and artificial feeds are fed. Fish should be on artificial feed when they reach about 2.5 cm in size. By the time striped bass are on artificial feed, the major losses associated with their culture have occurred and rearing the fish to a marketable size of 0.45 to 0.68 kg in 18 months is becoming more and more routine.

BLACK BASS

The largemouth bass (*Micropterus salmoides*) and the smallmouth bass (*Micropterus dolomieui*) are the only members of the black bass genus that are produced in large numbers. The largemouth bass is native to the midwestern and southeastern United States and northeastern Mexico. It has been introduced throughout the United States and many other countries worldwide.[14] The smallmouth bass was originally limited to the Midwest but has also been widely stocked across the United States and abroad.[15] In 1989, the U.S. Fish and Wildlife Service produced 3.2 million largemouth bass (3800 kg) and 630,000 smallmouth bass (1100 kg).[4] These fish were released in public waters. Many state governments also produce basses for stocking public waters. The market for black basses for stocking private waters is limited. In some states, it is illegal to sell black bass for direct human consumption. The only fish that can be sold are live bass for stocking purposes.

Large and smallmouth basses are usually spawned in earthen ponds.[13] Fry are removed from the spawning ponds and placed into fertilized rearing ponds. The fry consume zooplankton and other invertebrates in the ponds until they reach 2.5 to 5.0 cm in size, at which time they are released in public waters. Artificial feeds are not accepted by first-feeding basses, but 2.5 to 5.0 cm long basses can be trained to accept pelleted feed. Like striped bass, once on artificial feeds, black basses can be readily reared to advanced sizes.

Much of the work on *Morone* spp. and *Micropterus* spp. has been on their environmental and life history requirements and their acceptance of food. Little has been done with their nutritional requirements. This article reviews the nutritional requirements of the *Morone* spp. and *Micropterus* spp. and the various factors that influence their acceptance of food.

LABORATORY CULTURE

STRIPED AND HYBRID STRIPED BASSES

The striped bass, being anadromous, migrates into rivers in the spring to spawn. Hassler[16] has summarized the environmental requirements of striped bass. Striped bass eggs should be incubated around 18°C, at salinities of 1.5 to 3.0‰, and at dissolved oxygen levels greater than 5 ppm. Larvae are about 4 mm in length at hatching and their optimum salinity increases from around 3‰ during the first week after hatching to 33‰ during the sixth week.[16] The optimum salinity for post yolk-sac fry is 5 to 25‰, for juveniles it is 10 to 15‰, and 15 to 20‰ for adults. Optimum dissolved oxygen is 6 to 9 ppm.[16] First feeding should occur on about day 5 after hatching if the fish are held at 18°C. Striped bass, unlike other fish, do not appear to have a well-defined point of no return, when first feeding has to have occurred or the fish dies.[17] Eldridge et al.[17] found that starved striped bass could survive an average of 31 d. They did, however, find histological differences between fed and unfed larvae as early as 7 d posthatching.

A prepared feed for first-feeding striped bass on which they will survive has not been developed. Baragi and Lovell[18] reported that when striped bass larvae were offered live brine shrimp, heat-killed brine shrimp, and a prepared feed, the fish consumed all three diets, but mortality on the prepared diet was near 95% by day 16. Fry fed the prepared diet did not grow, whereas fish fed both brine shrimp treatments survived and grew at about the same rate. Enzymatic activity was measured in the guts of fish fed all three treatments from before first feeding through day 32 of the trial. Enzymes needed for the digestion of protein and starch, with the exception of pepsin, were present before feeding was initiated at levels sufficient to permit digestion of exogenous nutrients. The reason why the larvae survived on the brine shrimp but not on the prepared diet was not determined.

The amount of food required by first-feeding striped bass is a matter of debate. Chesney[19] found striped bass to grow well on brine shrimp at concentrations of 50 to 250 organisms per liter rearing volume. Eldridge et al.[17] reported that 100 organisms per liter were needed for normal growth to occur and over the range they studied, 0 to 5000 organisms per liter, as the number of organisms increased the rate of growth of the larvae increased. They also estimated the requirement for wild zooplankton to be about 1.31 times the requirement of brine shrimp.

Striped bass are cannibalistic as early as 6 d after hatching, even when brine shrimp are abundant.[20] If brine shrimp are not abundant, Paller and Lewis[21] found cannibalism and size variability to increase. The authors also found that striped bass fry could be successfully switched from brine shrimp to a formulated feed if the transition involved a weaning period.

One characteristic of striped bass that is rarely measured but can affect survival and growth is gas bladder inflation. Striped bass inflate their gas bladders between days 5 and 7 posthatch. If not inflated at that time, degeneration of tissue associated with the swim bladder begins and becomes evident 10 to 15 d posthatch.[22] Hadley et al.[23] reported that inflation of striped bass reared in intensive culture tanks is rarely above 50%. If this is true, the importance of recording the incidence of inflation in first-feeding studies of short duration is paramount if an accurate reflection of treatment effects is to be obtained.

BLACK BASS

Heidinger[24] and Coble[25] have reviewed the life history and biology of largemouth and smallmouth bass, respectively. Largemouth and smallmouth basses initiate spawning when water temperatures in the spring warms to 15 to 16°C. Largemouth bass spawn at temperatures as high as 24°C. Eggs of the largemouth bass hatch if incubated between 10 and 28°C. Smallmouth bass eggs hatch between 12 and 25°C. Normal incubation temperature range for both basses is between 18 and 23°C.

Largemouth and smallmouth basses complete their entire life cycles in freshwater. The addition of salt to concentrations equaling those of bass physiological salt concentrations is recommended when bass are transported or subjected to stress.[26]

At 22°C, eggs of largemouth bass take about 48 h to hatch and the larvae are about 4.6 mm in length. First feeding occurs when the fish are 5 d old and the fry become selective feeders when 7 d old.[27] Bass at first feeding to 12 mm in length consume zooplankton 0.2 to 0.4 mm in diameter.[27] Ten millimeter fry are able to consume a 1.0 mm carp (*Cyrpinus carpio*) egg, while 9 mm fry are not.[28] Several commercial and experimental feeds have been offered to largemouth bass fry, but survival has been generally less than 25%.[29,30] Fingerling largemouth bass, 2.5 to 3.8 cm long, can readily be trained to accept moist and semimoist feeds during intensive training periods.[31]

Eggs of the smallmouth bass at 23°C require 45 h to hatch, and the emerging larvae are 4.4 mm in length. The larvae are 9 mm in length when they start to feed.[32] It is safe to assume that first-feeding fry accept 0.2 to 0.5 mm in diameter zooplankton. Smallmouth bass fry accept commercial fry feeds more readily than largemouth bass. Ehrlich et al.[34] offered 11 mm fry a commercial feed, a combination of the commercial feed with *Artemia*, and *Artemia* alone. Best survival (85.3%) was obtained by feeding the commercial feed plus *Artemia*, while reasonably good success (39%) was obtained by feeding the commercial feed alone. Smallmouth can be trained to accept pelleted feed when they are in the 2.5 to 7.0 cm size range.[33] Cannibalism appears to be more of a problem with the black basses than with striped bass. Until black basses are trained to accept pelleted feed, care must be taken to provide them with an abundance of live food and to maintain uniform size fish within each rearing unit.

NUTRIENT REQUIREMENTS

STRIPED AND HYBRID STRIPED BASSES

Little is known about the nutrient requirements of striped and hybrid striped basses. Millikin[35,36] conducted a series of feeding trials with age-0 striped bass to determine the effect of dietary protein and lipid levels on growth. The best growth and protein utilization was obtained when 1.4 g fish were fed either a diet (Table 1) containing 47% protein and 12% lipid, or a diet containing 57% protein and 17% lipid. When the cost of each diet was taken into consideration, the lower protein and lipid diet was more economical to feed.[36] Berger and Halver[37] investigated the effects of protein, lipid, and carbohydrate levels on the growth of 9 to 16 g striped bass. They also found that striped bass performed well on high-protein and high-lipid diets. Maximum weight gain, feed efficiency, and protein efficiency ratio values were observed in fish fed a diet (Table 1) containing 52% protein and 13.2% fat (103.1 mg protein/kcal/g total energy).

Berger and Halver[37] also commented on the ability of striped bass to utilize lipids and carbohydrates. Their data indicated that striped bass were able to use lipids to spare the use of protein for energy at higher levels than other fish. They found no reduction in growth as dietary lipid levels increased up to 17% of the diet. They also found that striped bass were quite tolerant of dietary carbohydrate compared to other carnivorous fishes. As nitrogen-free extract increased from 15 to 23 to 33% of the diet, weight gain of the fish also increased.

Gallagher[38] conducted a feeding trial with striped bass × white perch (*Morone americana*)

TABLE 1
Experimental Diet Formulations for Striped Bass

Ingredient	Dietary protein and (lipid) level, % dry weight		
	47(12)[36]	57(17)[36]	52(13)[37]
Fish meal	39.6	48.4	69.9
Soy proteinate	21.6	26.4	
Dextrin			18.2
Corn starch	13.0	3.0	0.1
Cellulose	9.1	6.0	0.2
Fish oil	7.9[a]	11.3[a]	5.9[b]
Carboxymethyl cellulose			2.0
Calcium carbonate	1.5		
Sodium phosphate, dibasic	2.5		
Mineral mix	4.0[c]	4.0[c]	1.7[d]
Choline chloride	0.1	0.1	0.5
Vitamin mix	0.6[e]	0.6[e]	1.5[f]
Chromic oxide			0.5

[a] Menhaden oil supplemented with 200 mg ethoxyquin and 200 mg of DL-α-tocopheryl acetate per kg dry diet.
[b] Cod liver oil that contained 2500 IU of vitamin A per g and 250 IU of vitamin D per g.
[c] Bernhart-Tomarelli mineral mix.
[d] Natural Minerals Brand trace mineral mixture, Natural Minerals, Inc., Fullerton, CA.
[e] Vitamins, mixed with corn starch, fortified each kg of dry diet with 2000 IU vitamin A palmitate; 1000 IU vitamin D_3; 100 mg niacin; 100 mg ascorbic acid (ethylcellulose-coated); 300 mg inositol; 40 mg calcium pantothenate; 10 mg thiamine·HCl; 20 mg riboflavin; 40 mg menadione; 1 mg biotin; 20 μg vitamin B_{12}.
[f] Oregon Moist Pellet vitamin premix that contains the following as mg per kg premix: *d*-biotin, 40 mg; thiamine, 1576; riboflavin, 3527; pyridoxine, 1179; vitamin B_{12}, 4; niacin, 12566; ascorbic acid, 59524; vitamin E, 33510, vitamin K, 397; folacin, 849; myo-inositol, 17637; and *d*-pantothenic acid, 7055.

hybrids. Semipurified diets containing casein, egg white, corn starch, cellulose, cod liver oil, vitamins, and minerals were offered to juvenile fish. Fish fed the test diets grew at a slower rate that fish fed commercial trout feeds. Fish fed a diet containing 2.5 kcal/g of protein grew faster than fish fed a 3.5 kcal/g of protein diet. Fish fed both diets performed significantly better than fish fed a 4.5 kcal/g of protein diet.

The only research on the requirement of a specific nutrient by striped bass or one of its hybrids was by Webster.[39] He fed first-feeding striped bass brine shrimp nauplii from different sources. Brine shrimp from different sources contained different levels of palmitoleic acid, eicosapentaenoic acid (EPA), and linolenic acid. The level of EPA in brine shrimp affected survival and growth of first-feeding striped bass. Webster found that striped bass fry have a relatively high dietary requirement for EPA, perhaps 1.5% of the dry diet.

BLACK BASS

Anderson et al.[40] determined the minimum protein requirement for large- and smallmouth basses. Test fish at initiation of trials averaged 0.6 and 3.9 g for smallmouth bass, and 2.1 and 5.7 g for largemouth bass. Test diets contained fish protein concentrate, gelatin, dextrin, salmon oil, carboxymethyl cellulose, alphacel, and vitamin and mineral premixes. Protein and energy of the test diets ranged from 31.8% protein and 4.2 kcal/g of dry weight for gross energy to 68.2% protein and 4.8 kcal/g for energy. The minimum protein requirement for the smallmouth bass was found to be 45% on a dry weight basis and 40 to 41% for the largemouth bass.

It is apparent from the above discussion that only limited nutrient requirement information is available for striped and black basses. Generally, a species has to become important commercially before nutritional research is conducted. The hybrid striped bass is starting to generate some commercial interest. If this interest continues to increase, investigations on its nutritional requirements will also increase. The black basses will probably remain a sizable portion of the production on state and federal fish hatcheries in the Southeast, but their production as a food fish probably will not be in the near future.

PRACTICAL DIET FORMULATION

STRIPED AND HYBRID STRIPED BASSES

Prepared diets for first-feeding striped bass that produce normal growth and development are not available. Striped bass are most successfully started on live *Artemia* or zooplankton. A salmon starter ration is often offered to striped bass in conjunction with live food about 12 d after hatching.[38,41] Striped bass generally are not weaned from live food until they are 25 to 50 mm in length. Dry trout feeds are generally used for weaning (training) striped bass to pelleted feeds. Striped bass appear to more readily accept a semimoist pellet than a dry pellet. If trained on a semimoist pellet, striped bass can then be trained to accept a dry pellet. Klar and Parker[42] conducted three feeding trials with striped bass. They fed 2-month-old fish for 13 weeks, 7-month-old fish for 8 weeks, and a second group of 7-month-old fish for 11 months. They fed five different diets to the smallest fish and two diets to the larger fish. All diets were basically commercial diets. The authors final recommendation was to feed a high-quality trout feed, similar to the U.S. Fish and Wildlife Service trout grower ration GR-6-30. Information on practical striped-bass feeds is limited and until additional research is done the best approach is to feed high-quality commercial trout rations. Based on the results of several years of experience of feeding various-size striped bass at the National Fish Hatchery and Technology Center, guidelines for the proper food size and feeding rates of striped bass have been developed and are presented in Table 2.[43]

BLACK BASS

First-feeding black basses have not been successfully cultured on prepared feeds. Black basses are almost entirely reared on wild zooplankton in ponds. The addition of starter feeds to the tanks or ponds, as is done with striped bass, is not done with black bass. Ehrlich et al.[34] has successfully reared 11-mm smallmouth bass, which had been harvested from a pond, using a combination of live *Artemia* and a commercial feed. They obtained better growth on the combination of live and prepared feed than on either single food source. Ten millimeter largemouth bass that had been started on zooplankton have also been found to take carp eggs.[29]

Black basses are generally introduced to pelleted feeds when they are 3 to 5 cm in length.[44] Best results are obtained when basses are started on ground fish for a few days and then fed a semimoist trout feed. Once trained, bass can be readily converted to dry feeds.[44] For the past 10 years on the San Marcos National Fish Hatchery and Technology Center, black basses have been successfully trained, reared, and used as broodstock while being maintained on trout rations. Until further research is done, high-quality trout rations are the best diets available for rearing black bass. The use of semimoist feeds is also recommended to maintain the maximum number of fish on pelleted feed. Table 2 contains information on the proper feed sizes to use for largemouth bass.

Currently we are evaluating largemouth bass broodstock that have been fed either live fish or a pelleted feed.[45] We are looking at the effect of diet on annual lipid levels in various tissues,

TABLE 2
Food Sizes and Approximate Percentage of Body Weights to Feed to Striped[43] and Largemouth Bass[28,31]

Fish length (mm)	Food diameter (mm)		Percentage of body weight to be fed daily	
	Striped	Largemouth	Striped	Largemouth
Zooplankton				
5	0.2			
10	0.3	0.3		
20	0.4	0.4		
30	0.5	0.5		
Pelleted Feeds				
40	1.0	1.5	8.0	15.0
60	2.0	2.5	7.5	10.0
90	3.0	3.0	7.0	10.0
115	4.0	4.0	6.5	7.5
140	5.0	5.0	6.0	5.0
190	6.0	6.0	5.0	3.0
270	9.0	12.0	4.0	2.0
300	12.0	19.0	3.0	2.0

annual blood hormonal levels, as well as the effect on growth and gonadal development. We have not completed our evaluation but have observed significantly higher visceral fat levels in fish fed the pelleted feed.

It is surprising how little nutritional information is available on largemouth and striped basses, even though both species have been actively cultured on numerous state and federal hatcheries for close to 100 years. Perhaps the recent commercial interest in the black and striped basses will generate the funds needed to study these fishes.

REFERENCES

1. **Burgess, G. H.**, *Morone saxatilis* (Walbaum), striped bass, in *Atlas of North America Freshwater Fishes*, Lee, D. S., Gilbert, C. R., Hocutt, C. H., Jenkins, R. E., McAllister, D. E., and Stauffer, J. R., Jr., Eds., North Carolina State Museum of Natural History, Raleigh, NC, 1980, 576.
2. **Hassler, T. J.**, Species Profiles: Life Histories and Environmental Requirements of Coastal Fishes and Invertebrates (Pacific Southwest) — Striped Bass, U.S. Fish Wildl. Serv. Biol. Rep. 82(11.82), U.S. Army Corps of Engineers, TR EL-82-4, 1980, 29.
3. **Bishop, R. D.**, Evaluation of the striped bass (*Roccus saxatilis*) and white bass (*R. chrysops*) hybrids after two years, in *Proc. Annu. Conf. Southeast. Assoc. Game Fish Comm.*, 21, 245, 1968.
4. Fish and Fish Egg Distribution Report of the National Fish Hatchery System, Fiscal Year 1989/Report No. 24, 52, 1990.
5. **Dicks, M. and Harvey, D.**, Principal Contributors, Aquaculture Situation and Outlook Report, USDA, Economic Research Service, Aqua-3, Sept. 1989, 10.
6. **Harrell, R. M.**, Review of striped bass broodstock acquisition, spawning methods and fry production, in *The Aquaculture of Striped Bass: A Proceedings*, McCraren, J. P., Ed., UM-SG-MAT-84-01, Maryland Sea Grant Program, University of Maryland, College Park, MD, 1984, 45.

167

7. **Turner, C. J.**, Striped bass culture at Marion Fish Hatchery, in *The Aquaculture of Striped Bass: A Proceedings*, McCraren, J. P., Ed., UM-SG-MAT-84-01, Maryland Sea Grant Program, University of Maryland, College Park, MD, 1984, 59.
8. **Carlberg, J. M., Van Olst, J. C., Massingill, M. J., and Hovanec, T. A.**, Intensive culture of striped bass: a review of recent technological developments, in *The Aquaculture of Striped Bass: A Proceedings*, McCraren, J. P., Ed., UM-SG-MAT-84-01, Maryland Sea Grant Program, University of Maryland, College Park, MD, 1984, 89.
9. **Collins, C. M., Burton, G. L., and Schweinforth, R. L.**, High density culture of white bass × striped bass fingerlings in raceways using power plant heated effluent, in *The Aquaculture of Striped Bass: A Proceedings*, McCraren, J. P., Ed., UM-SG-MAT-84-01, Maryland Sea Grant Program, University of Maryland, College Park, MD, 1984, 129.
10. **Woods, L. C., III, Kerby, J. H., and Huish, M. T.**, Estuarine cage culture of hybrid striped bass, *J. World Maricult. Soc.*, 14, 595, 1983.
11. **Woods, L. C., III, Kerby, J. H., and Huish, M. T.**, Culture of hybrid striped bass to marketable size in circular tanks, *Prog. Fish-Cult.*, 47, 147, 1985.
12. **Stevens, R. C.**, Historical overview of striped bass culture and management, in *The Aquaculture of Striped Bass: A Proceedings*, McCraren, J. P., Ed., UM-SG-MAT-84-01, Maryland Sea Grant Program, University of Maryland, College Park, MD, 1984, 1.
13. **Dupree, H. K. and Huner, J. V.**, Propagation of black bass, sunfishes, tilapias, eels, and hobby fishes, in *Third Report to the Fish Farmers*, Dupree, H. K. and Huner, J. V., Eds., U.S. Fish and Wildlife Service, Washington, DC, 1984, chap. 9.
14. **Lee, D. S.**, *Micropterus salmoides* (Lacepede), largemouth bass, in *Atlas of North America Freshwater Fishes*, Lee, D. S., Gilbert, C. R., Hocutt, C. H., Jenkins, R. E., McAllister, D. E., and Stauffer, J. R., Jr., Eds., North Carolina State Museum of National History, Raleigh, NC, 1980-et seq. 608.
15. **Lee, D. S.**, *Micropterus dolomieui* (Lacepede), smallmouth bass, in *Atlas of North American Freshwater Fishes*, Lee, D. S., Gilbert, C. R., Hocutt, C. H., Jenkins, R. E., McAllister, D. E., and Stauffer, J. R., Jr., Eds., North Carolina State Museum of National History, Raleigh, NC, 1980-et seq. 605.
16. **Hassler, T. J.**, Species Profiles: Life Histories and Environmental Requirements of Coastal Fishes and Invertebrates (Pacific Southwest) — Striped Bass, U.S. Fish Wildl. Serv. Biol. Rep. 82 (11), U.S. Army Corps of Engineers, TR EL-82-4, 1988, 29.
17. **Eldridge, M. B., Whipple, J. A., Eng, D., Bowers, M. J., and Jarvis, B. M.**, Effects of food and feeding factors on laboratory-reared striped bass larvae, *Trans. Am. Fish. Soc.*, 110, 111, 1981.
18. **Baragi, V. and Lovell, R. T.**, Digestive enzyme activities in striped bass from first feeding through larva development, *Trans. Am. Fish. Soc.*, 115, 478, 1986.
19. **Chesney, E. J., Jr.**, Estimating the food requirements of striped bass larvae *Morone saxatilis*: effects of light, turbidity and turbulence, *Mar. Ecol. Prog. Ser.*, 53, 191, 1989.
20. **Braid, M. R.**, Incidence of cannibalism among striped bass fry in an intensive culture system, *Prog. Fish-Cult.*, 43, 210, 1981.
21. **Paller, M. H. and Lewis, W. M.**, Effects of diet on growth depensation and cannibalism among intensively cultured larval striped bass, *Prog. Fish-Cult.*, 49, 270, 1987.
22. **Doroshev, S. I. and Cornacchia, J. W.**, Initial swim bladder inflation in the larvae of *Tilapia mossambica* (Peters) and *Morone saxatilis* (Walbaum), *Aquaculture*, 16, 57, 1979.
23. **Hadley, C. G., Rust, M. B., Van Eenennaam, J. P., and Doroshov, S. I.**, Factors influencing initial swim bladder inflation by striped bass, *Am. Fish. Soc. Symp.*, 2, 164, 1987.
24. **Heidinger, R. C.**, Life history and biology of the largemouth bass, in *Black Bass Biology and Management*, Clepper, H., Ed., Sport Fishing Institute, Washington, D.C., 1975, 11.
25. **Coble, D. W.**, Smallmouth bass, in *Black Bass Biology and Management*, Clepper, H., Ed., Sport Fishing Institute, Washington, D.C., 1975, 21.
26. **Carmichael, G. J., Tomasso, J. R., Simco, B. A., and Davis, K. B.**, Characterization and alleviation of stress associated with hauling largemouth bass, *Trans. Am. Fish. Soc.*, 113, 778, 1984.
27. **Kauamara, G. and Washiyama, N.**, Ontogenetic changes in behavior and sense organ morphogenesis in largemouth bass and *Tilapia nilotica*, *Trans. Am. Fish. Soc.*, 118, 203, 1989.
28. **Brandt, T. M., Fries, J. N., Engeling, N. T., and Williamson, J. H.**, Status of the development of larval diets for black bass, presented at Aquaculture '89, Los Angeles, CA, Feb. 12 to 16, 1989, 3.
29. **Willis, D. W. and Flickinger, S. A.**, Intensive culture of largemouth bass fry, *Trans. Am. Fish. Soc.*, 110, 650, 1981.
30. **Brandt, T. M., Jones, R. M., Jr., and Anderson, R. J.**, Evaluation of prepared feeds and attractants for largemouth bass fry, *Prog. Fish-Cult.*, 49, 198, 1987.
31. **Brandt, T. M.**, Feeding largemouth bass, *Aquaculture*, 13(4), 63, 1987.
32. **Wallace, C. R.**, Embryonic and larval development of the smallmouth bass at 23°C, *Prog. Fish-Cult.*, 34, 237, 1972.

33. **Anderson, R. J.,** Feeding artificial diets to smallmouth bass, *Prog. Fish-Cult.*, 36, 145, 1974.
34. **Ehrlich, K. F., Cantin, M. C., Rust, M. B., and Grant, B.,** Growth and survival of larval and postlarval smallmouth bass fed a commercially prepared dry feed and/or *Artemia* nauplii, *J. World Aquacult. Soc.*, 20, 1, 1989.
35. **Millikin, M. R.,** Effects of dietary protein concentration on growth, feed efficiency, and body composition of age-0 striped bass, *Trans. Am. Fish. Soc.*, 111, 373, 1982.
36. **Millikin, M. R.,** Interactive effects of dietary protein and lipids on growth and protein utilization of age-0 striped bass, *Trans. Am. Fish. Soc.*, 112, 185, 1983.
37. **Berger, A. and Halver, J. E.,** Effect of dietary protein, lipid and carbohydrate content on the growth, feed efficiency and carcass composition of striped bass, *Morone saxatilis* (Walbaum), fingerling, *Aquacult. Fish. Manag.*, 18, 345, 1987.
38. **Gallagher, M.,** Nutritional requirements of striped bass and hybrids, in *Hybrid Striped Bass Culture; Status and Perspective*, Hodson, R., Smith, T., McVey, J., Harrell, R., and Davis, N., Eds., UNC Sea Grant College Program, North Carolina State University, Raleigh, NC, 1987, 53.
39. **Webster, C. D.,** Nutritional Value of Brine Shrimp Nauplii for Striped Bass Larvae, Ph.D. dissertation, Auburn University, Auburn, AL, 1989.
40. **Anderson, R. J., Kienholz, E. W., and Flickinger, S. A.,** Protein requirements of smallmouth bass and largemouth bass, *J. Nutr.*, 111, 1085, 1981.
41. **Lewis, W. M., Heidinger, R. C., and Tetzlaff, B. L.,** Tank culture of striped bass production manual, Illinois Striped Bass Project IDC F-26-R, Fisheries Research Laboratory, Southern Illinois University, Carbondale, IL, 1981, 115.
42. **Klar, G. T. and Parker, N. C.,** Evaluation of five commercially prepared diets for striped bass, *Prog. Fish-Cult.*, 51, 115, 1989.
43. **Graves, K. and Fries, J. N.,** unpublished data, 1987.
44. **Lovshin, L. L. and Rushin, J. H.,** Acceptance by largemouth bass fingerlings of pelleted feeds with a gustatory additive, *Prog. Fish-Cult.*, 51, 73, 1989.
45. **Rosenblum, P. M., Brandt, T. M., Mayes, K. E., and Chaterjee, N.,** Annual cycles of growth and reproduction in largemouth bass, *Micropterus salmoides, Am. Zool.*, 29, 97A, 1989.

TILAPIA, *OREOCHROMIS* spp.

Pierre Luquet

INTRODUCTION

World production of tilapia amounted to 280,000 tons in 1986, with figures as high as 500,000 tons also being reported.[1] These uncertainties are due to the difficulty of clearly distinguishing between the production derived from fish farming and that resulting from small-scale fisheries. Moreover, in some countries, such as Benin, very extensive forms of fish rearing exist, e.g., the acadja system, being similar to both fish farming and the exploitation of wild populations.

The main area of production is Asia (214,000 tons in 1978), the major producers being the Philippines, Taiwan, China, Indonesia, and Thailand.[2] In Africa, the continent from which the tilapia originated, the yearly production is quite low at 40,000 tons. In Israel, a country where the temperature is a limiting factor, production has increased regularly and represents 31% of the total production of pondfish.[3] This tendency to increase production is noticed in many countries, and spectacular results have been seen. In Latin America, there is no actual intensive tilapia culture, however, the management and exploitation of reservoirs allow for significant production.

The main cultured species are *Oreochromis niloticus, O. aureus, O. hornorum, O. mossambicus,* and *Tilapia zillii,* as either pure or hybrid strains. It should be pointed out that referring to pure stock is quite relative with respect to tilapia, as numerous uncontrolled minglings have been carried out, and it is often more proper to refer to them as *Oreochromis* sp. or as a strain coming from a certain country or a certain fish farm. For example, the red tilapia performs differently in seawater whether it comes from Florida or Taiwan.[4]

At least with the female mouthbrooders of the *Oreochromis* genus, there exists a very marked delay in the growth of females, and monosex culture of males is preferred. Thus, it is very important to indicate the sex of the cultured fish, and whether sex control was obtained through manual sorting, sex reversal by hormone treatment, or hybridization.

LABORATORY CULTURE

The assessment of dietary requirements implies the lack of any limiting or masking factors that, through a growth impairment, may lead to a misevaluation of requirements and recommendations.

The following experimental conditions have been suggested to be optimal for *O. niloticus:*[5] temperature, 28 to 31°C; dissolved oxygen, above 3 mg/l; ammonia nitrogen, below 2.5 mg/1; and biomass, 0.4 to 0.6 kg/l. An example of a purified test diet that has given excellent performance for *O. niloticus* is presented in Table 1. Because of their continuous feeding behavior and small stomach capacity, tilapia should be fed more frequently than most other fish. The number of feedings per day should be nine for juvenile and six for 100 g or larger *O. niloticus* and eight for juvenile, and three for adult *O. mossambicus.*

NUTRIENT REQUIREMENTS

PROTEINS AND AMINO ACIDS

The quantitative protein requirements of *O. niloticus, O. mossambicus, O. aureus,* and *T. zillii* have recently been reviewed by De Silva et al.[7] and are listed in Table 2. Some additional

TABLE 1
An Example of a Purified Diet for Tilapia

Ingredient	Percent
Casein	33
α-Starch	20
Dextrin	27
Corn oil	8
Mineral mix	5
Vitamin mix	2
Cellulose	5
Nutrient content	
Moisture	9.2
Crude ash	5.3
Crude protein	29.0
Crude carbohydrate	50
Crude lipid	8.1
Gross energy	450 kcal/100 g

From Wang, K.-W., Takeuchi, T., and Watanabe, T., *Bull. Jpn. Soc. Sci. Fish,* 51, 141, 1985. With permission.

data for *O. niloticus* or its hybrids (*O. hornorum* × *O. niloticus, O. niloticus* × *O. aureus*) indicate an optimal requirement ranging from 28 to 32% crude protein.[22,23] Summing up the above data, it appears that tilapia have a protein requirement of 30 to 35% of the diet.

Water salinity appears to influence the protein requirement, being lower (20 to 24%) at full salinity, as indicated in Table 3.[10,24-26] The general trend of variations with the age, size, and feeding rate remains to be better documented.[6,27-29]

Under practical conditions, two additional considerations have lead to lower recommended dietary protein levels. First, the most economical dietary protein content, even if it does not support maximum growth, is close to 28%.[7] In ponds, fish may have access to natural food that is rich in protein, thus dietary protein levels as low as 20 to 25% have been estimated to be adequate.[30-32]

Only limited information is available on the essential amino acid requirements of tilapia. Research in this area encounters various methodological problems, e.g., a lower growth rate is usually observed when fish are fed diets based on crystalline amino acids, even if pH or mineral adjustments are made.[33-36] The mechanism for this reduced growth rate is not known, however, in at least one case this type of test diet performed quite well.[36,37] For *O. mossambicus*, Jackson and Capper[34] have estimated the lysine (1.62% diet), methionine (0.53%), and arginine (1.52%) requirements using practical diets containing 40% crude protein. The quantitative requirements for the 10 essential amino acids have been determined by growth experiments on *O. niloticus* by Santiago and Lovell,[37] and these data are compared in Table 4 with proposed values for *O. mossambicus* based on indirect methods.[35]

Most of the studies involved in the determination of the quantitative protein requirements have used high-quality protein sources, such as fish meal or casein. Additional studies have been done in the laboratory to evaluate cheaper feedstuffs as alternative sources of dietary protein for practical diets. In general, animal byproduct meals have received little attention, however, blood meal, poultry byproduct meal, and feather meal can be used if incorporation levels higher than 10% are avoided.[38,39]

In general, oilseed meals are used quite well by tilapia, even if some contradictory results have been observed. Soybean meals, either as the conventional soybean meal or as the full-fat soybean meal, if properly processed to destroy the trypsin inhibitor activity, have been shown to be a very satisfactory feedstuff. Animal-protein free feeds containing from 50 to 68% soybean meal can be used if dicalcium phosphate (3%) is added.[40,41] Cottonseed meal is a

TABLE 2
The Dietary Protein Requirements of Tilapia[a]

Species/weight	% protein	Ref.
O. niloticus		
Fry to 5 g	30	8–11
	35	12,13
	38–47	14
O. mossambicus		
0.5–1.0 g	40	15
3–4 cm	29–38	16
12–70 g	30	17
O. aureus		
0.3–0.5 g	36	18
2.5–7.5 g	34	19
T. zillii		
1.4–1.8 g	35	20

[a] Adapted from De Silva et al.[7]

TABLE 3
Effects the Salinity on the Protein Requirement of Tilapia

Species	Salinity ‰	% protein	Ref.
O. niloticus			
	0	30.4	10
	5	30.4	10
	10	28.0	10
	15	28.0	10
	29	30.0	24
O. niloticus × *O. aureus*	32–34	24.0	25
O. urolepsis hornorum ×			
O. mossambicus	37	20.0	26

TABLE 4
Quantitative Essential Amino Acid Requirement for Tilapia[a]

Amino acid	*O. mossambicus*[b]	*O. niloticus*[c]
Arginine	2.82	4.20
Histidine	1.05	1.72
Isoleucine	2.01	3.11
Leucine	3.40	3.39
Lysine	3.78	5.12
Methionine	0.99	2.68
Methionine + cystine		3.21
Phenylalanine	2.50	3.75
Phenylalanine + tyrosine		5.54
Threonine	2.93	3.75
Tryptophan	0.43	1.00

[a] Expressed as a percent of dietary protein.
[b] Data based on carcass composition and daily deposition.[35]
[c] Data resulting from growth experiments.[37]

From Santiago, C. B. and Lovell, R. T., *J. Nutr.*, 118, 1540, 1988. With permission.

TABLE 5
Protein Digestibility of Selected Feedstuffs
for Tilapia[15,25,29,46,47]

Feedstuff	Percent
Fish meal	72–92
Poultry offal meal	53
Soybean meal	91–94
Groundnut meal	79
Wheat midlings	20–76
Corn	53–83
Brewers grains	62–63

TABLE 6
Essential Fatty Acid Requirements of Tilapia

Species	Requirement	Ref.
T. zillii	1% 18:2n-6 or	52
	1% 22:4n-6	
O. niloticus	1% 18:2n-6	53
	1% 22:4n-6	
	0.5% 18:2n-6	54
O. aureus	n-6 series	49

more controversial feedstuff, due to its content of gossypol, a known antinutritional factor. Dietary inclusion levels of 20 to 30% cottonseed meal have been reported to be both safe and useful.[39,42-45] Inadequate research has been carried out on the potential use of copra, groundnut, sunflower, and rapeseed meals to recommend levels higher than 30% be used in practical diets.[44] The potential use of unconventional feedstuffs, such as single-cell proteins, alga, and leguminous vegetables, also needs additional study.[29] Selected data on protein digestibility of several feedstuffs for tilapia are given in Table 5.

LIPIDS AND ESSENTIAL FATTY ACIDS

Information on lipid and essential fatty acid (EFA) requirements for tilapia is not as complete as for coldwater fishes.[48] There is no doubt that essential fatty acids are required, even though fat-free diets have occasionally given similar growth as fortified diets.[49] The energetic function of lipids has not been well demonstrated, because incorporation rates up to 5% do not always improve experimental or practical feeds.[50,51] Dietary lipid levels of 5 to 6% are often used.

The EFA requirement for n-6 rather than n-3 fatty acids seems well established for tilapia, except perhaps for *O. aureus*. The results concerning this species are so conflicting that they need to be confirmed. *T. zillii* require either linolenic acid (18:2n-6) or 22:4n-6.[52] The EFA requirements of tilapia are summarized in Table 6. Corn oil, soybean oil, and safflower oil are suitable lipid sources to provide for the EFA requirements of tilapia.[48,55]

CARBOHYDRATES

Tilapia appear to utilize carbohydrates quite well. Digestibility values as high as 75 to 79% have been reported for wheat flour at a dietary level of 35%, and 50 to 56% for potato starch at a dietary level as high as 85%.[56] Dietary levels as high as 40% are well tolerated and are used to support growth by *O. niloticus* and *T. zillii*.[57,58] Dried cassava (or manioc or tapioca), a starch-rich root, is also well utilized at dietary inclusion levels ranging from 30 to 60%.[11,59]

VITAMINS

Vitamin requirements of tilapia have not been widely studied. The only available information concerns vitamins C, E, and B_{12}. Additional studies have been concerned with the physiological activity of the digestive flora and its ability to synthesize some group B vitamins, thus providing for part of the needs.

In *O. mossambicus*, the absence of any detectable activity of the enzyme L-gulonolactone oxidase was suggested to indicate the essentiality of L-ascorbic acid.[60] The quantitative requirement for normal growth of *T. aurea* has been reported to be 50 mg/kg diet.[61] Deficient tilapia performed poorly, and the gross deficiency signs included mild scoliosis and occasional hemorrhages of the fins, mouth, and swim bladder. In *O. niloticus*, five different forms of ascorbic acid have been tested and found to perform equally well as dietary sources in terms of growth, feed conversion, and protein utilization.[62] The five forms included L-ascorbic acid, the sodium salt of L-ascorbic acid, glyceride-coated L-ascorbic acid, the barium salt of L-ascorbic acid 2-sulfate, and ascorbyl palmitate. Supplemental ascorbic acid in broodstock fish (*O. mossambicus*) feed improved hatchability and fry condition. Fry produced from broodstock fed an unsupplemented diet performed poorly with respect to survival, growth, and feed utilization, whereas fry resulting from broodstock fed a fortified diet performed better.[63]

The vitamin E requirement of *O. niloticus* was found to be 5 to 10 mg of DL-α-tocopheryl acetate/100 g of a diet containing 5% lipid.[51] Feeding diets containing 10 to 15% lipid increase the α-tocopherol requirement to 50 mg of DL-α-tocopheryl acetate/100 g diet.

Gastrointestinal bacterial flora counts as high as 10^9 to 10^{10}/g were found in *O. niloticus* reared in either fresh water or seawater.[64,65] It has been shown that such a high level of bacteria may induce changes in the structure of the intestinal epithelium and may lead to biosynthetic activity in the different digestive compartments. Such is the case with tilapia, which have a long intestine. Intestinal synthesis of vitamin B_{12} has been demonstrated in *O. niloticus*. This synthesis was significantly reduced when either a cobalt-free diet or a diet containing an antibiotic was fed. These workers concluded that vitamin B_{12} supplementation is not necessary for normal growth and erythropoiesis in tilapia.[30,66]

Vitamin B_{12}, as well as other B vitamins (such as biotin and thiamine), can be synthesized by various forms of freshwater phytoplankton. Since most tilapia are microphagous and omnivorous feeders, phytoplankton can serve as a significant source of B vitamins.[67]

MINERALS

As with other fish, mineral requirement studies in tilapia are very difficult to conduct, thus only limited information is available. For example, *O. mossambicus*, an euryhaline fish, is able to take up calcium efficiently via the gills and phosphorus via the gut.[68] It is likely that the trace elements are also taken up from the rearing water, even in fresh water. This excellent ion and osmoregulating fish can grow in fresh water with Ca^{2+} levels as low as 0.2 mM. Since significant levels of both macro and trace elements can be readily absorbed from the rearing water, experimentally controlling the total dietary intake of a certain mineral is difficult.

Since various feedstuffs, such as fish meal, contain a certain level of minerals, mineral premixes are often deleted from practical diets. Nevertheless, based on studies of the availability of certain minerals in such feedstuffs and the possible interactions between them and phytates, the use of such premixes is recommended.[69,70]

Dietary mineral requirements are summarized in Table 7. The calcium requirement is higher than that reported for other warmwater fish, such as the channel catfish. Such a difference may be related to the presence or absence of scales. Calcium deficiency leads to a reduction of growth, feed efficiency, and bone mineralization (calcium, phosphorus, and total ash) in *O. aureus*.[71] Phosphorus deficiency also leads to a reduction in growth and bone mineralization. The requirement for available phosphorus is about the same as those reported for other fish species. The availability of tricalcium phosphate contained in fish meal was

TABLE 7
Mineral Requirements of Tilapia

Species/mineral	Requirement	Remark	Ref.
O. aureus			
Calcium	<6.5 g/kg	In calcium-free water	71
O. niloticus			
Phosphorus	9.0 g/kg	Available P	72
Magnesium	0.59–0.77 g/kg	Acetate form	73
Zinc	10 mg/kg		69
Manganese	12 mg/kg		69
Copper	3–4 mg/kg		69

TABLE 8
Composition of Tilapia Feeds Used in Israel

Ingredient	Fish meal	Standard
Fishmeal	25	15
Soybean meal	—	20
Wheat, ground	20	20
Sorghum, ground	55	45
Protein	25.0	25.0
Fat	4.0	3.0
Ash	6.0	5.0
Calcium	0.8	0.6
Phosphorus	0.7	0.7

Note: For *O. aureus* hybrids, in intensive cage or pond culture. Both diets give similar results for growth and feed conversion. Supplemental oil does not improve growth.

From Viola, S. and Ariel, Y., *Bamidgeh*, 35, 9, 1983. With permission.

found to be quite good (65%).[72] The magnesium requirement is also about the same as for other fish. Magnesium acetate appears to be a better source than magnesium oxide or magnesium sulfate. The requirement appeared to increase with increasing levels of dietary protein.[73,74] Due to the similarities of these requirements with those of rainbow trout, a similar mineral supplementation can be used.

PRACTICAL DIET FORMULATION AND FEEDING

Research findings on nutrient requirements should normally be applied to practical feed formulations. But in the case of tilapia, economic constraints and local feedstuff availability are often the major factors considered in feed formulation. In the same way, if the natural productivity of the pond is not known, it is usually taken into account as far as its qualitative contribution is concerned. Thus, vitamin supplementation is often omitted or has not led to any convincing improvement.

The practical formulations commonly in use are very simple and only consist of three or four feedstuffs, mainly of plant origin. Examples of some formulas that are currently being used in commercial production systems are presented in Tables 8 and 9.

In ponds, feeding pelleted feeds results in no conclusive advantage over meal feeds from a zootechnical point of view. On the contrary, pelleted feeds are necessary in cage culture to reduce the loss of uneaten feed. Recommended sizes of crumbles and pellets are listed in Table 10.[27] For cage culture of fingerlings (from 5 to 10 g onwards), an alternative method of feeding involves feeding a meal feed using feeding rings.[77,78] Feeding frequency and daily feeding

TABLE 9
Composition of Tilapia Feeds Used in Niger

Ingredient	Broodstock	Fingerlings	Food size fish
Groundnut meal	50	40	45
Wheat or rice bran	50	40	50
Fishmeal	—	20	5
Crude protein	31	37	31.5

Note: For *O. niloticus*, vitamin supplementation did not improve growth. Pelleted form for food size fish reared in cages only.

From Parrel, P., Ali, I., and Lazard, J., *Bois et Forêts des Tropiques*, 212, 71, 1986. With permission.

TABLE 10
Recommended Pellet Sizes for Juvenile and Adult Tilapia

Fish size/age	Particule size (diameter)
Fry: first 24 h	Liquify
Fry: 2nd–10th day	0.5 mm
Fry: 10th–30th day	0.5–1 mm
Fry: 30th day–juveniles	0.5–1.5 mm
0.5–1.0 g	
1–30 g	1–2 mm
20–120 g	2 mm
100–250 g	3 mm
250 g +	4 mm

From Jauncey, K. and Ross, B., *A Guide to Tilapia Feeds and Feeding*, Institute of Aquaculture, University of Stirling, Scotland, 1982. With permission.

TABLE 11
Feeding Table for Tilapia Fingerlings in Ponds and Cages

Fish size (g)	Feeding rate[a]		Meals/d	
	Pond	Cage	Pond	Cage
0–5	15–7	—	6	—
5–10	6.6	12	5	4
10–15	5.3	10	4	4
15–20	5.3	8	3	4
20–30	4.6	6	2	4

[a] Expressed as percent of body weight/d.

From Parrel, P., Ali, I., and Lazard, J., *Bois et Forêts des Tropiques*, 212, 71, 1986. With permission.

rates are still empirically adjusted, and depend on the availability of devices in order to prevent or correct falls in levels of dissolved oxygen. Examples of recommended feedings rates are presented in Tables 11 and 12. It should be noted that the values for feeding in cages are higher than those for feeding in ponds to take into account the availability of natural food. Feeding rates are also affected by water temperature, as indicated in Table 13. Overfeeding at low temperatures (below 20°C) increases mortality. Feeding should be stopped at water temperatures below 16°C.

TABLE 12
Feeding Table for Tilapia with Pellets Containing 25% Protein in Growing Ponds

Fish size (g)	Monoculture	Polyculture with carp
5–10	0.5	0.4
10–20	0.8	0.6
20–50	1.6	1.3
50–70	2.0	1.6
70–100	2.4	1.9
100–150	2.7	2.2
150–200	3.0	2.5
200–300	3.7	3.0
300–400	4.5	3.6
400–500	5.2	4.2
500–600	6.0	4.8

Note: Expressed as grams/fish/d.

From Marek, M., *Bamidgeh*, 27, 57, 1975. With permission.

TABLE 13
Feeding Rate and Number of Daily Meals as Modified by Water Temperature

Temperature	B.W.< 100 g		B.W. > 100 g	
	% B.W./d	No./d	% B.W./d	No./d
t > 24°	100% NR	4	100% NR	4
24° > t > 22°	70% NR	3	50% NR	2
22° > t > 20°	50% NR	2	40% NR	2
20° > t > 18°	35% NR	2	25% NR	2
18° > t > 16°	20% NR	2	10% NR	1
16° > t	No feeding		No feeding	

Note: Feeding rate is expressed as a percent of normal rate (NR).

From Parrel, P., Ali, I., and Lazard, J., *Bois et Forêts des Tropiques*, 212, 71, 1986. With permission.

REFERENCES

1. **FAO**, Aquaculture production (1984–1986), FAO Fisheries Circular, 815, FIDI/C815, 1989.
2. **Pillay, T. V. R.**, Asian aquaculture: an overview, in *Aquaculture in Asia*, Mohan Joseph, M., Ed., Asian Fisheries Society, Indian Branch, Mangalore, 1990, 31.
3. **Sarig, S.**, The fish culture industry in Israel in 1988, *Israeli J. Aquaculture — Bamidgeh*, 41, 50, 1989.
4. **Ernst, D. H., Ellingson, L. J., Olla, B., Wicklund, R. I., Watanabe, W. O., and Grover, J. J.**, Production of Florida red tilapia in seawater pools: nursery rearing with chicken manure and growout with prepared feed, *Aquaculture*, 80, 247, 1989.
5. **Mélard, C.**, Les bases biologiques de l'élevage intensif du tilapia du Nil, *Cahiers d'Écologie Appliquée* 6, 1, 1986.
6. **Wang, K-W., Takeuchi, T., and Watanabe, T.**, Optimum protein and digestible energy levels in diets for *Tilapia nilotica*, *Bull. Jpn. Soc. Sci. Fish.* 51, 141, 1985.

7. **De Silva, S., Gunasekera, R. M., and Atapattu, D.**, The dietary protein requirements of young tilapia and an evaluation of the least cost dietary protein levels, *Aquaculture*, 80, 271, 1989.

8. **Appler, H. N. and Jauncey, K.**, The utilization of a filamentous green alga [*Cladophora glomerata* (L) Kuztin] as a protein source in pelleted feeds for *Sarotherodon* (*Tilapia*) *niloticus* fingerlings, *Aquaculture*, 30, 21, 1983.

9. **Appler, H. N.**, Evaluation of *Hydrodicton reticulatum* as a protein source in feeds for *Oreochromis* (*Tilapia*) *niloticus* and *Tilapia zillii*, *J. Fish. Biol.*, 27, 327, 1985.

10. **De Silva, S. S. and Perera, M. K.**, Effects of dietary protein level on growth, food conversion, and protein level on growth, food conversion, and protein use in young *Tilapia nilotica* at four salinities, *Trans. Am. Fish. Soc.*, 114, 584, 1985.

11. **Wee, K. L. and Ng, L. T.**, Use of cassava as an energy source in pelleted feed for the tilapia *Oreochromis niloticus* L., *Aquacult. Fish. Manage.*, 17, 129, 1986.

12. **Santiago, C. B., Banes-Aldaba, M., and Laron, M. A.**, Dietary crude protein requirement of *Tilapia nilotica* fry, *Kalikasan Philip, J. Biol.*, 11, 255, 1982.

13. **Teshima, S., Kanazawa, A., and Uchiyama, Y.**, Optimum protein levels in casein-gelatin diets for *Tilapia nilotica* fingerlings, *Mem. Fac. Fish., Kagoshima Univ.*, 34, 45, 1985.

14. **Tacon, A. G. J., Jauncey, K., Falayne, A., Pantha, M., Macgoven, I., and Stafford, E. A.**, The use of meat and bone meal, hydrolyzed feathermeal and soybean in practical fry and fingerling diets for *Oreochromis niloticus*, in *Proc. Int. Symp. on Tilapia in Aquaculture*, Filshelson, L. and Yaron, Z., Eds., Tel Aviv University, Tel Aviv, Israel, 1983, 336.

15. **Jauncey, A.**, The effects of varying dietary protein level on the growth food conversion, protein utilization and body composition of juvenile tilapias (*Sarotherodon mossambicus*), *Aquaculture*, 27, 43, 1982.

16. **Cruz, E. M. and Laudencia, I. L.**, Protein requirements of *Tilapia mossambica* fingerlings, *Kalikasan Philipp. J. Biol.*, 6, 177, 1977.

17. **Jackson, A. J., Capper, B. S., and Matty, A. J.**, Evaluation of some plant proteins in complete diets for the tilapia (*Sarotherodon mossambicus*), *Aquaculture*, 27, 97, 1982.

18. **Davis, A. T. and Stickney, R. R.**, Growth response of *Tilapia aurea* to dietary protein quality and quantity, *Trans. Am. Fish. Soc.*, 107, 479, 1978.

19. **Winfree, R. A. and Stickney, R. R.**, Effects of dietary protein and energy on growth, feed conversion efficiency and body composition of *Tilapia aurea*, *J. Nutr.*, 111, 1001, 1981.

20. **Mazid, M. A., Tanaka, Y., Katayama, T., Rahma, M. A., Simpson, K. L., and Chichester, C. O.**, Growth response of *Tilapia zillii* fingerlings fed isocaloric diets with variable protein levels, *Aquaculture*, 18, 115, 1979.

21. **Siddiqui, A. Q., Howlader, M. S., and Adam, A. A.**, Effects of dietary protein levels on growth, feed conversion and protein utilization in fry and young nile tilapia, *Oreochromis niloticus*, *Aquaculture*, 70, 63, 1988.

22. **Yong, W.-Y., Takeuchi, T., and Watanabe, T.**, Relation between digestible energy contents and optimum energy to protein ratio in *Oreochromis niloticus* diet, *Bull. Jpn. Soc. Sci. Fish.*, 55, 869, 1989.

23. **Shiau, S.-Y., Chuang J.-L, and Sun, C.-L.**, Inclusion of soybean meal in tilapia (*Oreochromis niloticus* × *O. aureus*) diets at two protein levels, *Aquaculture*, 65, 251, 1987.

24. **Fineman-Kalio, A. S. and Camacho, A. S.**, The effect of supplemental feeds containing different protein:energy ratios on the growth and survival of *Oreochromis niloticus* (L) in brackishwater ponds, *Aquacult. Fish. Manage.*, 18, 139, 1987.

25. **Shiau, S.-Y. and Huang, S.-L.**, Optimal dietary protein level for hybrid tilapia (*Oreochromis niloticus* × *O. aureus*) reared in seawater, *Aquaculture*, 81, 119, 1989.

26. **Clark, A. E., Watanabe, W. O., Olla, B. L., and Wicklund, R. I.**, Growth, feed conversion and protein utilization of Florida red tilapia fed isocaloric diets with different protein levels in seawater pools, *Aquaculture*, 88, 75, 1990.

27. **Jauncey, K. and Ross, B.**, *A Guide to Tilapia Feeds and Feeding*, Institute of Aquaculture, University of Stirling, Scotland, 1982.

28. **Wang, K., Takeuchi, T., and Watanabe, T.**, Effect of dietary protein levels in diets on growth of *Tilapia nilotica*, *Bull. Jpn. Soc. Sci. Fish.*, 51, 133, 1985.

29. **Luquet, P.**, Practical considerations on the protein nutrition and feeding of tilapia, *Aquat. Liv. Resour.*, 2, 99, 1989.

30. **Lovell, R. T.**, Feeding tilapia, *Aquaculture*, 7, 42, 1980.

31. **Wanninga, N. D., Weerakoon, D. E. M., and Muthukumarana, G.**, Cage culture of *S. niloticus* in Sri Lanka: effect of stocking density and dietary crude protein levels on growth, in *Finfish Nutrition in Asia. Methodological Approaches to Research and Development*, Cho, Y., Cowey, C. B., and Watanabe, T., Eds., Ottawa, Ont., IDRC-233e, 1985, 113.

32. **Newman, M. W., Huezo, H. E., and Hughes, D. G.**, The response of all-male tilapia hybrids to four levels of protein in isocaloric diets, in *Proc. World Maricul. Soc.*, 10, 788, 1979.

33. **Mazid, M. A., Tanaka, Y., Katayama, T., Simpson, K. L., and Chichester, C. O.**, Metabolism of amino acids in aquatic animals. III. Indispensable amino acids for *Tilapia zillii*, *Bull. Jpn. Soc. Sci. Fish.*, 44, 739, 1978.

34. **Jackson, A. J. and Capper, B. S.**, Investigation into the requirements of tilapia *Sarotherodon mossambicus* for dietary methionine, lysine and arginine in semi-synthetic diets, *Aquaculture*, 29, 289, 1982.

35. **Jauncey, K., Tacon, A. G. J., and Jackson, A. J.**, The quantitative essential amino acid requirements of *Oreochromis (Sarotherodon) mossambicus*, in *Proc. Int. Symp. on Tilapia in Aquaculture*, Fishelson, L. and Yaron, Z., Eds., Tel Aviv University, Tel Aviv, Israel, 1983, 328.

36. **Yamada, S., Tanaka, Y., Katayama, T., Sameshima, M., and Simpson, K. L.**, Plasma amino acid changes in *Tilapia nilotica* fed a casein and a corresponding free amino acid diet, *Bull. Jpn. Soc. Sci. Fish.*, 48, 1783, 1982.

37. **Santiago, C. B. and Lovell, R. T.**, Amino acid requirements for growth of nile tilapia, *J. Nutr.*, 118, 1540, 1988.

38. **Otubusin, S. O.**, Effects of different levels of blood meal in pelleted feeds on tilapia, *Oreochromis niloticus*, production in floating bamboo net-cages, *Aquaculture*, 65, 263, 1987.

39. **Viola, S. and Zohar, G.**, Nutrition studies with market size hybrids of tilapia (*Oreochromis*) in intensive culture. 3. Protein levels and sources, *Bamidgeh*, 36, 3, 1984.

40. **Viola, S., Arieli, Y., and Zohar, G.**, Animal-protein-free feeds for hybrid tilapia (*Oreochromis niloticus* × *O. aureus*) in intensive culture, *Aquaculture*, 75, 115, 1988.

41. **Wee, K. L. and Shu, S.-W.**, The nutritive value of boiled full-fat soybean in pelleted feed for nile tilapia, *Aquaculture*, 81, 303, 1989.

42. **Ofojekwu, P. C. and Ejike, C.**, Growth response and feed utilization in the tropical cichlid *Oreochromis niloticus* (Linn.) fed on cottonseed-based artificial diets, *Aquaculture*, 42, 27, 1984.

43. **Robinson, E. H., Rawles, S. D., Oldenburg, P. W., and Stickney, R. R.**, Effects of feeding glandless or glanded cottonseed products and gossypol to *Tilapia aurea*, *Aquaculture*, 38, 145, 1984.

44. **Jackson, A. J., Capper, B. S., and Matty, A. J.**, Evaluation of some plant proteins in complete diets for the tilapia *Sarotherodon mossambicus*, *Aquaculture*, 27, 97, 1982.

45. **El-Sayed, A. R. M.**, Long-term evaluation of cottonseed meal as a protein source for nile tilapia, *Oreochromis niloticus* (Linn.), *Aquaculture*, 84, 315, 1990.

46. **Hanley, F.**, The digestibility of foodstuffs and the effects of feeding selectivity on digestibility determinations in tilapia, *Oreochromis niloticus (L)*, *Aquaculture*, 66, 163, 1987.

47. **De Silva, S. S. and Perera, M. K.**, Digestibility in *Sarotherodon niloticus* fry: effect of dietary protein level and salinity with further observations on variability in daily digestibility, *Aquaculture*, 38, 293, 1984.

48. **Stickney, R. R. and Hardy, R.**, Lipid requirements of some warmwater species, *Aquaculture*, 79, 145, 1989.

49. **Stickney, R. R. and McGeachin, R. B.**, Response of *Tilapia aurea* to semipurified diets of differing fatty acid composition, in *Proc. Int. Symp. on Tilapia in Aquaculture*, Fishelson, L. and Yaron, Z., Eds., Tel Aviv University, Tel Aviv, Israel, 1983, 346.

50. **Viola, S. and Arieli, Y.**, Nutrition studies with tilapia hybrids. 2. The effects of oil supplements to practical diets for intensive aquaculture, *Bamidgeh*, 35, 44, 1983.

51. **Satoh, S., Takeuchi, T., and Watanabe, T.**, Requirement of *Tilapia* for α-tocopherol, *Nippon Suisan Gakkaishi*, 53, 119, 1987.

52. **Kanazawa, A., Teshima, S., Sakamoto, M., and Awal, M. A.**, Requirments of *Tilapia zillii* for essential fatty acids, *Bull. Jpn. Soc. Sci. Fish.*, 46, 1353, 1980.

53. **Teshima, S., Kanazawa, A., and Sakamoto, M.**, Essential fatty acids of *Tilapia nilotica*, *Mem. Fac. Fish., Kagoshima Univ.*, 31, 201, 1982.

54. **Takeuchi, T., Satoh, S., and Watanabe, T.**, Requirement of *Tilapia nilotica* for essential fatty acids, *Bull. Jpn. Soc. Sci. Fish.*, 49, 1127, 1983.

55. **Takeuchi, T., Satoh, S., and Watanabe, T.**, Dietary lipids suitable for the practical feed of *Tilapia nilotica*, *Bull. Jpn. Soc. Sci. Fish.*, 49, 1361, 1983.

56. **Barash, H., Neumark, H., Heffer, E., Iosif, B., and Itzkovich, J.**, An improved technology for determining diet digestibility by tilapia, in *Proc. Int. Symp. on Tilapia in Aquaculture*, Fishelson, L. and Yaron, Z., Eds., Tel Aviv University, Tel Aviv, Israel, 1983, 338.

57. **El-Sayed, A. F. M. and Garling, D. L.**, Carbohydrate-to-lipid ratios in diets for *Tilapia zillii* fingerlings, *Aquaculture*, 73, 157, 1988.

58. **Anderson, J., Jackson, A. J., Matty, A. J., and Capper, B. S.**, Effects of dietary carbohydrate and fiber on the tilapia *Oreochromis niloticus* (Linn.), *Aquaculture*, 37, 303, 1984.

59. **Viola, S., Arieli, Y., and Zohar, G.**, Unusual feedstuffs (tapioca and lupin) as ingredients for carp and tilapia feeds in intensive culture, *Israeli J. Aquaculture (Bamidgeh)*, 40, 29, 1988.

60. **Soliman, A. K., Jauncey, K., and Roberts, R. J.**, Qualitative and quantitative identification of L-gulonolactone oxidase activity in some teleosts, *Aquacult. Fish. Manage.*, 16, 249, 1985.

61. **Stickeny, R. R., McGeachin, R. B., Lewis, D. H., Marks, J., Sis, R. F., Robinson, E. H., and Warts, W.**, Response of *Tilapia aurea* to dietary vitamin C, *J. World Maricul. Soc.*, 15, 179, 1984.

62. **Soliman, A. K., Jauncey, K., and Roberts, R. J.**, The effect of varying forms of dietary ascorbic acid on the nutrition of juvenile tilapias (*Oreochromis niloticus*), *Aquaculture*, 52, 1, 1986.
63. **Soliman, A. K., Jauncey, K., and Roberts, R. J.**, The effect of dietary ascorbic acid supplementation on hatchability, survival rate and fry performance in *Oreochromis mossambicus* (Peters), *Aquaculture*, 59, 197, 1986.
64. **Sugita, H., Ishida, Y., Deguchi, Y., and Kadota, H.**, Bacterial flora in the gastrointestine of *Tilapia nilotica* adapted to seawater, *Bull. Jpn. Soc. Sci. Fish.*, 48, 987, 1982.
65. **Sugita, H., Tonkuyama, K., and Deguchi, Y.**, The intestinal microflora of carp *Cyprinus carpio*, grass carp *Ctenopharynqodon idella*, and tilapia *Sarotherodon niloticus*, *Bull. Jpn. Soc. Sci. Fish.*, 51, 1325, 1985.
66. **Lovell, R. T. and Limsuwan, T.**, Intestinal synthesis and dietary nonessentiality of vitamin B_{12} for *Tilapia nilotica*, *Trans. Am. Fish. Soc.*, 111, 485, 1982.
67. **Nishijima, T., Shiozaki, R., and Hata, Y.**, Production of vitamin B_{12}, thiamine, and biotin by freshwater phytoplankton, *Bull. Jpn. Soc. Sci. Fish.*, 45, 199, 1979.
68. **Flik, G., Wendele Bonga, S. E., Kolar, Z., Mayer-Gostan, N., and Fenwick, J. C.**, Environmental effects on Ca^{2+}-uptake in the cichlid teleost *Oreochromis mossambicus*, in *Abstr. 7th Conf. European Soc. Comp. Physiol. Biochem.*, Barcelone, August 26–28, B 6.4, 1985.
69. **Watanabe, T., Satoh, S., and Takeuchi, T.**, Availability of minerals in fish meal to fish, *Asian Fish. Sci.*, 1, 175, 1988.
70. **Satoh, S., Poe, W. E., and Wilson, R. P.**, Effect of supplemental phytate and/or tricalcium phosphate on weight gain, feed efficiency and zinc content in vertebrae of channel catfish, *Aquaculture*, 80, 155, 1989.
71. **Robinson, E. H., Rawles, S. D., Yette, H. E., and Greene, L. W.**, An estimate of the dietary calcium requirements of fingerling *Tilapia aurea* reared in calcium free water, *Aquaculture*, 41, 389, 1984.
72. **Watanabe, T., Takeuchi, T., Murakami, A., and Ogino, C.**, The availability to *Tilapia nilotica* of phosphorus in white fish meal, *Bull. Jpn. Soc. Sci. Fish.*, 46, 897, 1980.
73. **Dabrowska, H., Günther, K. D., and Meyer-Burgdorff, K.**, Availability of various magnesium compounds to tilapia (*Oreochromis niloticus*), *Aquaculture*, 76, 269, 1989.
74. **Dabrowska, H., Meyer-Burgdorff, K., and Günter, K. D.**, Interaction between dietary protein and magnesium level in tilapia (*Oreochromis niloticus*), *Aquaculture*, 76, 277, 1989.
75. **Viola, S. and Arieli, Y.**, Nutrition studies with tilapia (*Sarotherodon*). 1. Replacement of fishmeal by soybean meal in feeds for intensive tilapia culture, *Bamidgeh*, 35, 9, 1983.
76. **Parrel, P., Ali, I., and Lazard, J.**, Le développement de l'aquaculture au Niger: un exemple de'élevage de *Tilapia* en zone sahéliene, *Bois et Forêts des Tropiques*, 212, 71, 1986.
77. **Lazard, J.**, L'élevage du *Tilapia* en afrique; données techniques sur sa pisciculture en étang, *Bois et Forêts des Tropiques*, 206, 33, 1984.
78. **Miller, J. W.**, A preliminary study of feeding pelleted versus non-pelleted feeds to *Tilapia nilotica* L. in ponds, in *Finfish Nutrition and Fishfeed Technology*, Vol. 1, Halver, J. E. and Tiews, K., Eds., Heenemann, Berlin, 1979.
79. **Marek, M.**, Revision of supplementary feeding tables for pond fish, *Bamidgeh*, 27, 57, 1975.

YELLOWTAIL, *SERIOLA QUINQUERADIATA*

Sadao Shimeno

INTRODUCTION

In Japan, yellowtail is one of the most popular and important fish and its cultured production has been the highest of the cultured fish species for more than 20 years. Yellowtail culture has shown rapid growth since the 1960s, due to increased demand and improved culture techniques, with tonnage and value of its production reaching 150,000 tons and 1 billion dollars, respectively, in 1980. Since then, however, growth in these figures has slowed down, and the number of culture farms has also shown a decrease. This has been due to the deterioration of environmental water quality originating from expanding production and to static low fish prices.

Yellowtail is mainly cultured in the southwest districts of Japan (the regions of Kyushu and Shikoku, and the southern side of Honshu), especially in Ehime and Kagoshima Prefectures. Yellowtail fingerlings are collected in a southwest sea region of Japan, then enclosed in floating-type or sinker-type net cages (100 to 400 m²) set in bays, and reared using frozen fish, such as sand lance, anchovy, and mackerel, as their main feed.[1] Recently, however, the Oregon moist pellet and other formulated feeds are gradually becoming popular, in order to maintain water quality and improve meat quality. Yellowtail of 1 to 5 g body weight at enclosing time in May grow to a size of 1 to 2 kg in December of the same year. Most of the fish are harvested in December or January of the next year, but some may be reared longer.

LABORATORY CULTURE

Since yellowtail are quite active and grow rapidly, it is preferable to use a large rearing aquarium to conduct nutritional experiments. One should not use a closed-type circulating aquarium because the high oxygen consumption rate of the fish requires a continuous supply of a large amount of seawater. In our laboratory, yellowtail juveniles (5 to 50 g) are reared in a 0.5 to 1.0 ton flow-through fiber-reinforced plastic tank with sufficient supply of seawater and aeration in a well-lighted room.

Juvenile yellowtail do not feed well on a tough dry pellet or a moist casein diet without feeding stimulants. Takeda[2] and Hosokawa et al.[3] have investigated the effects of moisture content and feeding stimulants in the diet. These workers found that (1) a soft moist pellet having about 40% moisture was preferred, (2) supplementation with several amino acids improved the nutritive value of casein,[4,5] and (3) dietary addition of inosinic acid, alanine, and proline increased the feeding activity of the fish.[6] Based on these results, a moist pellet-type purified casein diet is commonly used for nutritional studies (Table 1). After mixing the powdered materials, oil, and water, the pH of the mixture is adjusted to 6.5 with a NaOH solution, an appropriate size of moist pellet is prepared using a pellet mill, and the diet is stored in a freezer. The pH adjustment is important for obtaining good growth and high feed efficiency. Since juvenile yellowtail grow rapidly due to their high feeding rate, it is necessary to feed more than two times a day, but also it is possible to evaluate an experimental result by rearing the fish for only 30 to 40 days.

When the body weight is over 50 g, a semipurified diet mixed with fish meal, instead of casein, is also used, and fish of this size can be reared in a small floating net cage in a bay.[7]

TABLE 1
Composition of a Purified Test Diet
for Yellowtail[3,4]

Ingredient[a]	Percent
Vitamin-free casein	64.0
L-Isoleucine	0.5
L-Lysine·HCl	1.0
L-Methionine	0.5
L-Cystine	0.5
L-Arginine·HCl	2.0
L-Histidine·HCl·H$_2$O	0.5
Pollack liver oil[b]	14.0
White dextrin	3.0
Mineral mixture[44]	8.0
Water-soluble vitamin mixture[c]	2.0
Fat-soluble vitamin mixture[d]	1.0
L-Proline	0.131
L-Alanine	0.086
5'-Inosinic acid	0.153
Carboxymethyl cellulose	2.63

[a] These ingredients are thoroughly mixed with an equivalent amount of water and finally adjusted to pH 6.5 with 25% NaOH solution.

[b] The residual oil obtained by distilling away vitamin A from pollack liver oil.

[c] The mixture (mg/2 g): thiamine HCl, 6.0; riboflavin, 20.0; pyridoxine HCl, 4.0; Ca-pantothenate, 28.0; nicotinic acid, 80.0; biotin, 0.6; folic acid, 1.5; cyanocobalamin, 0.06; choline chloride, 800; inositol, 400; ascorbic acid, 200; para-aminobenzoic acid, 40.0; cellulose, 419.84.

[d] The mixture (mg/1 g): vitamin A acetate, 0.688; calciferol, 0.0045; α-tocopherol, 40.0; menadione, 4.0; pollack residual oil, 955.3075.

NUTRIENT REQUIREMENTS

Compared with other fish species, yellowtail have a low ability to utilize carbohydrates and a high ability to utilize protein and lipid because of their carnivorous nature.[8] As the vitamin and mineral requirements of the fish are also slightly different from those in other fishes, these nutrient requirements of yellowtail are described below.

PROTEIN

According to many reports on protein digestibility in yellowtail, the digestibility of fresh and frozen fish proteins, in general, is high (90 to 95%), and those of fish meal proteins are also high (80 to 90%). Unlike carp, protein digestibility in yellowtail is affected by the carbohydrate content in the diet. For example, as the gelatinized starch content in the diet increased successively from 0 to 10, 20, and 40%, the protein digestibility decreased from 84 to 82, 78, and 56%, respectively. Also, diarrhea-like feces was observed in fish fed the high carbohydrate diets, indicating an adverse effect of dietary carbohydrate on protein digestibility.[9,10] Despite many reports of similar effects,[11] there is a report that found that the protein digestibility was constant and independent of the carbohydrate content of the diet.[12]

Changes in the plasma amino acid concentration of yellowtail after feeding vary slightly depending on the type and composition of the diet, but in general, the plasma levels reach a maximum in 2 to 4 h after feeding and then decrease rapidly.[13] Such changes are remarkably quick compared to rainbow trout, indicating that protein digestion and the subsequent absorption and metabolism of amino acids are rapid in yellowtail.

Takeda et al.[14] and Shimeno et al.[7] have examined the effects of dietary protein content and calorie to protein ratio (C/P ratio) in yellowtail. Fish of 100 g average body weight were fed fish meal diets containing different levels of protein and lipid for 30 d. Similar or superior growth and feed efficiency of fish fed a high-protein diet were observed in fish fed diets containing 53 to 55% of protein and about a 70 C/P ratio. Good results were also obtained in fish fed a diet with 56% protein and about a 70 C/P ratio, which contained 9% lipid with different levels of protein and carbohydrate.[9] Similar results using purified casein diets have been reported by Deshimaru et al.[15] These results suggest that the minimum dietary protein content needed to produce maximum growth is about 50%, which is higher than those in diets of omnivorous and herbivorous fishes, reflecting the feeding habit of the fish. Regarding the protein source for the diet, few studies have been made, except for a few reports showing that brown fish meal has similar nutritive value as white fish meal.[16,17] Recently, studies on a partial replacement of fish meal with soybean meal have been started.[18] The amino acid requirements have not been determined.

CARBOHYDRATE

It is desirable to add inexpensive carbohydrates to fish feed as an energy source, because dietary protein can be saved and will consequently lower the feed cost. However, yellowtail have a low ability to utilize dietary carbohydrate, and a high dietary intake of carbohydrate may cause metabolic disturbance.[8]

Shimeno et al.[9,10] reared young yellowtail using fish meal diets in which the gelatinized starch content varied from 0 to 40%, and reported that the growth rate, efficiencies of feed, and energy, as well as body fat and glycogen content, were higher in fish fed 10% and 20% starch than those fed 0% and 40% starch. An examination of the effect of the ratio of carbohydrate to lipid in the diet showed that the best results were obtained in the group fed the diet containing 10% of each nutrient, compared with either of the two other groups fed diets containing 20% of one of the nutrients.[19] Additional studies on the optimum dietary ratio of the two nutrients, using six diets containing different amounts of carbohydrate and lipid but containing the same amount of protein and energy, indicated that the growth and feed conversion were highest in fish fed a diet containing 53% protein, 14% lipid, and 8% carbohydrate.[20]

Similar results have been obtained from experiments using casein diets,[15] and the digestibility values for carbohydrate and protein were fairly high in low-carbohydrate containing diets.[9,12] Furthermore, Shimeno et al.[9,10] have investigated the effects of dietary carbohydrate on the hepatic enzymes and the body composition of yellowtail. Dietary carbohydrate markedly increased the activities of phosphofructokinase (PFK), phosphoglucose isomerase (PGI), glucose-6-phosphate dehydrogenase (G6PDH), and phosphogluconate dehydrogenase (PGDH), as well as the contents of glycogen and fat in the liver, and also lowered the activities of glucose-6-phosphatase (G6Pase) and fructose-1,6-diphosphatase (FDPase). These findings seem to indicate that in fish fed diets containing limited amounts of carbohydrate, glycolysis and lipogenesis are enhanced, and gluconeogenesis and amino acid degradation are inhibited, as shown in Figure 1 (upper). It is evident from the metabolic data, in addition to data such as the growth, feed conversion, digestibility, and glucose tolerance, that yellowtail can adapt to low carbohydrate diets. But the fish cannot adapt, unlike carp, to high carbohydrate diets, because these diets caused extremely low digestibilities of both carbohydrate and protein, a low glucose tolerance, and an abnormal carbohydrate metabolism, resulting in growth retardation.[9]

With regard to the carbohydrate source, Furuichi et al.[21,22] reported that gelatinized starch

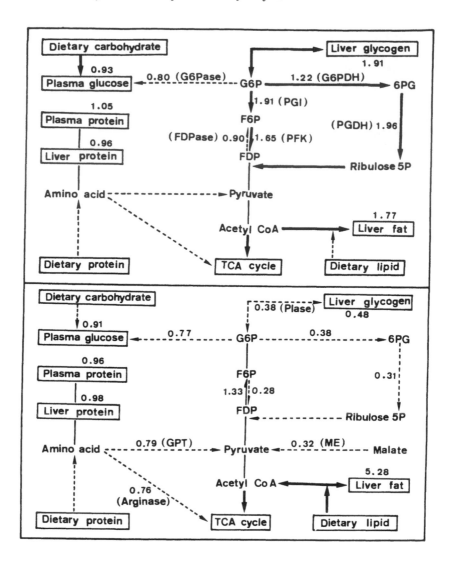

FIGURE 1. Metabolic adaptation to dietary carbohydrate[9,10] (upper) and lipid[29] (lower) in young yellowtail. Values are expressed as a ratio of the activity, content, and concentration in fish fed a high-carbohydrate and a high-lipid diet, to those in fish fed a high-protein control diet, respectively.

was more effectively utilized for growth by yellowtail juveniles than glucose when fed for 30 d in casein diets containing 10 or 20% of either carbohydrate. Similar results have been found using fish meal diets.[23] Subsequently, these workers investigated the cause of this difference with respect to the postfeeding changes in carbohydrate absorption and carbohydrate-metabolizing enzyme activities.[21] They concluded that, since dietary glucose is absorbed quickly before the increment of enzyme activities, most of the glucose is not metabolized but excreted, resulting in low utilization, whereas the gelatinized starch is digested slowly and absorbed after the peak of the enzyme activities, resulting in a high utilization. Deshimaru et al.[24] also compared the nutritive values of several carbohydrates and showed that (1) raw starch was the best source, followed by gelatinized starch, with dextrin being the most inferior source tested; and (2) the nutritive value of raw starch in potato, sweet potato, and rice varied depending on the source. This report is of great interest, because it shows that a high molecular weight starch is more effective as the carbohydrate source for yellowtail feed than low molecular weight carbohydrates, such as dextrin and glucose, and that a cheaper carbohydrate, such as raw starch, is more effective than gelatinized starch. These results will also be useful for the development and improvement of practical feeds.

Although no specific study on the carbohydrate requirement has been carried out, growth and feed conversions were better for fish fed diets containing low levels of carbohydrate than those fed a carbohydrate-free control diet.[9,10] In addition, carbohydrates serve as the least expensive source of dietary energy and as a binder for pelleting. Therefore, a limited amount of carbohydrate should be added to yellowtail diets.[10]

LIPID

Yellowtail utilize lipids effectively and the various lipids commonly used in fish feeds are highly digestible (90 to 99%), independent of lipid type, content, diet form, or water temperature.[25] It has been reported that, when young yellowtail were fed for 30 d on fish meal diets containing 0 to 19% of lipid, the growth and feed efficiency increased as the amount of dietary lipid increased and reached a maximum in the 11% lipid group. The efficiency and retention of dietary protein were also improved, resulting in a protein-sparing effect of lipid.[7] Takeda et al.[14] reported similar results, but Deshimaru et al.[15] found, using casein diets, that the optimum dietary lipid content varied depending on the source, e.g., when squid liver oil and pollack liver oil were added to the diet, maximum growth and feed efficiency were observed in fish fed 9 and 15% lipid, respectively. They have also compared the effects of various oils at a 9% level and showed that the nutritive value of squid liver oil, skip-jack oil, and sardine oil, each being rich in n-3 highly unsaturated fatty acid (HUFA) content, were higher than that of pollack liver oil and herring oil, both of which have lower HUFA content.[15,26] It had previously been established that the nutritive value of corn oil and soybean oil was lower than that of pollack liver oil for yellowtail, unlike some freshwater fish, such as carp.[27] Yone[28] suggested that yellowtail require HUFAs, such as eicosapentaenoic acid (EPA) and docosahexaenoic acid (DHA), as essential fatty acids, while many freshwater fish require linoleic acid and linolenic acid. In addition, studies on the effects of the quantity and quality of HUFAs revealed that (1) the growth rate and feed efficiency increased as the dietary HUFA content increased and reached a maximum at a level of 1.6 to 2.1%;[15] and (2) the nutritive value of the two HUFAs (EPA and DHA) was similar.[24]

With regard to the metabolic response to dietary lipid, the analysis of enzyme activities and body composition of young yellowtail fed high-lipid diets seems to show that dietary lipid causes a decrease in glycolysis, lipogenesis, and amino acid degradation, as well as an increase in assimilation and preservation of the nutrient as an energy source[29] (Figure 1, lower). These metabolic responses result in the protein-sparing effect of lipid.

Feeding oxidized lipid to yellowtail caused growth depression, muscular dystrophy, and myopathy, whereas dietary supplementation of α-tocopherol effectively prevented these abnormalities.[30-32] It should be noted that the storage and transportation of lipid sources must be conducted carefully, because oxidation of lipid is enhanced by light and heat.

VITAMINS

During the last 10 years, a series of studies on the vitamin requirements in yellowtail juveniles has been carried out in our laboratory. Information has been obtained on the essential vitamins, vitamin deficiency signs, vitamin requirements, recommended vitamin mixtures, and so on.

Initially, Hosokawa et al.[33,34] studied the essentiality of each vitamin for yellowtail using a purified casein diet, as shown in Table 1, and showed that the fish require a total of 11 water-soluble vitamins (thiamine, riboflavin, pyridoxine, pantothenic acid, nicotinic acid, biotin, folic acid, cyanocobalamin, choline, inositol and ascorbic acid) and 2 fat-soluble vitamins (calciferol and α-tocopherol). A deficiency in any vitamin resulted in an initial loss in appetite, followed by reduced growth rate, and finally death as a common sequence, but many other

TABLE 2
Vitamin Deficiency Signs in Yellowtail[33,34]

Vitamin	Time in days to develop signs	Deficiency signs[a]
Thiamine	24	Dark coloration, congestion in fins and opercles
Riboflavin	30	Congestion in fins and eyes, dark coloration, cloudy lens
Pyridoxine	14	Epileptiform fits, convulsions, ataxia
Pantothenic acid	10	Blue coloration, clubbed gills, rapid death
Nicotinic acid	20	Skin lesions, hemorrhage in body surface, loss of caudal fin
Biotin	24	Hemorrhage in stomach and intestine, ataxia
Folic acid	24	Congestion in fins and opercles, dark coloration, macrocytic anemia
Cyanocobalamin	25	Congestion in fins and opercles, hemorrhagic liver, macrocytic anemia
Choline	3	Dark coloration, poor food conversion
Inositol	20	Dark coloration, ataxia, skin lesions
Ascorbic acid	20	Scoliosis, dark coloration, hemorrhage in body surface, hypochromic anemia
PABA[b]	—	None detected
Vitamin A	19	Dark coloration, ataxia, hemorrhage in fins and eyes, hemorrhagic liver, anemia, poor growth of opercula
Calciferol	—	None detected
α-Tocopherol	28	Dark coloration, ataxia, congestion in fins and opercula, convulsion, muscular dystrophy
Menadione	—	None detected

[a] Anorexia, reduced weight gain, and mortality are common vitamin deficiency signs and thus are not included in the table.
[b] Para-aminobenzoic acid.

deficiency signs specific for each vitamin were observed, as listed in Table 2. These deficiency signs are similar to those observed in other fish species, but the time of onset of the deficiency and death are sooner than with most other species, indicating that yellowtail may have a higher sensitivity to vitamin deficiency. Such a property is thought to result from a rapid utilization of the vitamins in juveniles that maintain active nutrient metabolism. However, the period between the feeding of a vitamin-deficient diet and the termination of growth varies depending on each vitamin. For example, the reduced growth rate was observed soon after feeding diets deficient in choline, pantothenic acid, and pyridoxine (3 to 14 d), while the period was longer in fish fed diets deficient in the other water- and fat-soluble vitamins.[34] Fish fed diets devoid of para-amino benzoic acid, calciferol, and menadione had similar growth rates as the controls after 50 d of feeding, with no apparent deficiency signs nor increase in mortality. Therefore, these vitamins seem to be unnecessary for the fish or necessary only in trace amounts.

Secondly, Hosokawa et al.[35-37] investigated the quantitative requirements of each essential vitamin for yellowtail. They fed juveniles for 30 to 50 d on diets containing graded levels of a particular vitamin in an otherwise complete, purified diet; compared the weight gain, hepatic vitamin content, and so on; and calculated the minimum requirement for the dietary vitamin level that results in normal growth and hepatic vitamin content and no deficiency signs. Figure 2 shows the effects of dietary pantothenic acid on weight gain and liver pantothenic acid content. As is evident from the figure, the minimum requirement of pantothenic acid based on the liver vitamin content (35.9 mg/kg dry diet) is considerably higher than that based on weight gain (13.5 mg). Such a difference is commonly observed with other vitamins, indicating that liver vitamin content responds more rapidly to dietary vitamin intake than weight gain.

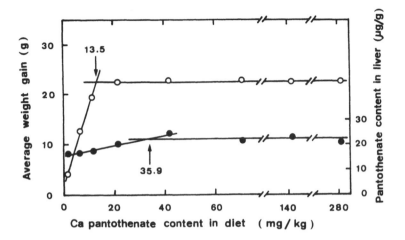

FIGURE 2. Relationships of calcium pantothenate content in diet to average weight gain (○) and pantothenate content in liver (●) of yellowtail fingerlings.[36]

TABLE 3
Vitamin Requirements of Yellowtail[35-38]

Vitamin	Requirement[a] (mg/kg dry diet)	Recommended (mg/kg dry diet)
Thiamine	11.2	22.4
Riboflavin	11.0	22.0
Pyridoxine	11.7	23.4
Pantothenic acid	35.9	71.8
Nicotinic acid	12.0	96.0
Biotin	0.67	1.34
Folic acid	1.2	2.4
Cyanocobalamin	0.053	0.424
Choline	2920.0	5840.0
Inositol	423.0	846.0
Ascorbic acid	122.0	1950.0
Para-aminobenzoic acid	N[b]	
Vitamin A	5.68	6.88
Calciferol	N[b]	
α-Tocopherol	119.0	595.0
Menadione	N[b]	

[a] Minimum levels required in the diet to prevent deficiency signs.
[b] No dietary requirement demonstrated under test conditions.

Table 3 shows the requirement of each vitamin for yellowtail juveniles, based on the liver vitamin content for all vitamins except cyanocobalamin and ascorbic acid, which are based on weight gain.[35-37] The requirement values for yellowtail are similar to those for carp, but differ from those for chinook salmon: the values for cyanocobalamin, vitamin A, and α-tocopherol are considerably higher in yellowtail than in chinook salmon. Several vitamins are know to be synthesized by intestinal microorganisms in carp, but there is no knowledge for yellowtail. Thus, it is unclear whether the fish obtain any appreciable amount of vitamins from microbial synthesis in the gastrointestinal tract.

Since the quantitative requirement of one vitamin varies depending on the amount of other coexisting vitamins, a vitamin mixture prepared on the basis of the minimum requirement level of each vitamin does not necessarily meet the total vitamin requirement of the fish.

Feeding experiments indicated that adding a vitamin mixture based on each requirement level resulted in reduced growth and induced various deficiency signs, but normal growth was obtained when another vitamin mixture was used in which the content of each vitamin was increased to two to eight times higher than the requirement level.[38]

Because many vitamins are involved in specific nutrient metabolism as coenzymes, Hosokawa et al.[40,41] also studied the effects of dietary protein and fat levels on vitamin requirements. The pyridoxine requirement increased 1.4-fold when the protein level in the diet was increased from 59 to 79%. When the lipid level in the diet was increased from 8 to 15 and 25%, the α-tocopherol requirement increased markedly (from 35 to 93 and 160 mg/kg diet), while the pyridoxine requirement increased only slightly (from 3.0 to 3.4 and 3.5 mg). Based on these results and those discussed above, they proposed the recommended level of each vitamin to be added to yellowtail diets, as presented in Table 3.[38]

The efficacy of large additions of vitamins to diets has been confirmed not only in relation to weight gain, but also from the viewpoint of immunopotentiation effects in yellowtail. For example, the phagocytic activity and antibody titers were reported to be improved in fish when 2 to 5 and 16 to 50 times the quantitative requirement of α-tocopherol and ascorbic acid, respectively, were added to a diet, resulting in an increased resistance to pseudotuberculosis and streptcoccicosis.[39]

In order to diagnose the vitamin status of fish, Hosokawa et al.[42] compared response rates of the blood vitamin content, hepatic enzyme activity, metabolite content, etc., to dietary vitamin levels. Blood vitamin content was the most sensitive diagnostic indicator for thiamine, riboflavin, pyridoxine, and α-tocopherol status in yellowtail.

MINERALS

Since fish can absorb minerals from both the feed and culture water, mineral requirements are difficult to determine, particularly in the case of yellowtail, which are cultured in seawater, which is rich in minerals, thus detailed studies on mineral requirements have been slow to develop.

Shimizu[43] examined the absorption pathway of minerals by performing tracer experiments using radioisotopes and demonstrated that Fe, Zn, Sb, and Sr were absorbed chiefly from the diet, Cr mainly from seawater, and Mn and Co from both.

Recent studies in our laboratory on mineral requirements in yellowtail fed a purified casein diet have indicated the following. The optimum level of a dietary mineral mixture prepared in accordance with Halver[44] was found to be 8%.[45] Fish fed a diet without the mineral mixture showed an initial loss in appetite, followed by dark coloration, scoliosis, anemia, and death, but these signs were improved by feeding a recovery diet containing the mineral mixture.[46] In omission tests, some of the deficiency signs described above were observed in fish fed diets devoid of Fe and P, showing the essentiality of these two minerals.[46] The minimum level of dietary Fe and P was estimated to be about 6 to 16 and 670 mg/100 g dry diet, respectively, although these values must be evaluated further in detail. On the other hand, good results, equivalent to the control group fed diets containing the complete mineral mixture, were observed in fish fed diets devoid in Na, K, or Cl, with no apparent deficiency signs. Fish fed diets devoid in Ca and Mg showed slightly reduced growth and feed efficiency at the later stage of the feeding period, although no deficiency signs were observed.[46]

Feeding studies using fish meal diets that contain considerably high levels of minerals have indicated the following. The optimum level of Halver's mineral mixture was found to be 1%.[47] It was necessary to add Fe and P sources to this diet.[16,17] An iron-proteinate (a commercial compound binding Fe to soybean protein hydrolysate) was found to be superior to iron citrate or iron sulfate as the Fe source.[48,49]

PRACTICAL DIET FORMULATION

Until around 1985, frozen fish, such as sardine, mackerel, sand lance, etc., were used as the main feed for yellowtail culture in Japan. However, with the increase in total culture and subsequent pollution of environmental water, growth retardation, diseases, and subsequent mortality became frequent. In addition, the meat quality of cultured fish has been criticized by consumers. In order to solve these problems, practical formulated feeds, such as the Oregon moist pellet (OMP), single moist pellet (SMP), and extruded pellet (EP), have been studied and developed, and have recently been used in certain districts.[50]

OMP is a moist pellet prepared with a pellet mill from raw materials comprising a powdered formulated diet, minced fish, and oil (10:10:1, in general). Recently, OMP has come into considerable use in various districts. OMP is readily accepted by yellowtail, is convenient for controlling the quality and quantity of vitamins and oil to be added to the feed, and good powdered formulated diets are available on the market for preparing OMP. This pellet, however, has the disadvantage of requiring facilities and labor for its preparation and storage.

SMP is also a moist pellet prepared with a pellet mill from a mixture comprised of a powdered formulated diet, fresh water, and oil (10:4:1, in general). This pellet has the same advantages as those of OMP. In addition, this pellet has stable nutritive values and is hygienic, because it does not require fresh fish as a raw material. However, SMP has not been put into practical use, since it cannot be actively fed in low temperature seasons.

EP is a soft and dry pellet prepared using an extruder, and is a new type of feed that has been put on the market for yellowtail culture since 1988 in Japan. This pellet has the advantage of requiring no facilities for its preparation and storage, and is now used in certain districts.

Each of these three feeds has its own advantages and disadvantages, and accordingly, culturists select the best one by taking the feed price and regional characteristics into consideration. Although these formulated feeds, except for OMP, have not become popular in yellowtail culture, the author wishes to develop and popularize complete formulated feeds as quickly as possible in order to keep the culture environment clean and to produce healthy and high-quality fish.

SPECIAL DIETS

Recently, artificial spawning and rearing of yellowtail larvae have been successfully accomplished in Japan, and live feeds, such as rotifers, *Artemia* nauplii, *Tigriopus*, and marine copepoda, are fed to the larvae.[1] However, it is essential to put a larval formulated diet into practical use for the efficient production of yellowtail seedlings. As a candidate for this feed, a microparticulate diet has been studied.[51] This diet, however, is not currently used because it cannot be used as the sole feed source to maintain satisfactory larval growth.

REFERENCES

1. **Watanabe, T.**, Intensive marine farming in Japan, in *Intensive Fish Farming*, Shepherd, J. and Bromage, N., Eds., BSP Professional Books, Oxford, 1988, 239.
2. **Takeda, M.**, Nutrition and diet in yellowtail-1. Adequate moisture content and feeding stimulants in the diet, *Fish Culture*, 22(3), 98, 1985.
3. **Hosokawa, H., Sumita, Y., and Takeda, M.**, Effects of Dietary Cellulose Levels on the Growth, Feed Conversion and Proximate Body Composition of Yellowtail Fingerlings, Reports of Usa Marine Biological Institute, No. 1, 9, 1979.
4. **Tachi, T.**, Study on a Purified Test Diet for Yellowtail Fingerlings, M.S. thesis, Kochi University, Nankoku, Japan, 1975.

5. **Deshimaru, O., Kuroki, K., and Yone, Y.**, Suitable levels of lipids and ursodesoxycholic acid in diet for yellowtail, *Nippon Suisan Gakkaishi*, 48, 1265, 1982.
6. **Takii, K.**, Survey of Feeding Stimulants for Yellowtail Fingerlings, M.S. thesis, Kochi University, Nankoku, Japan, 1977.
7. **Shimeno, S., Hosokawa, H., Takeda, M., and Kajiyama, H.**, Effects of calorie to protein ratios in formulated diet on the growth, feed conversion and body composition of young yellowtail, *Nippon Suisan Gakkaishi*, 46, 1083, 1980.
8. **Shimeno, S., Hosokawa, H., Hirata, H., and Takeda, M.**, Comparative studies on carbohydrate metabolism of yellowtail and carp, *Nippon Suisan Gakkaishi*, 43, 213, 1977.
9. **Shimeno, S.**, *Studies on Carbohydrate Metabolism in Fish*, A. A. Balkema, Rotterdam, 1982, 1.
10. **Shimeno, S., Hosokawa, H., and Takeda, M.**, The importance of carbohydrate in the diet of a carnivorous fish, in *Finfish Nutrition and Fishfeed Technology*, Vol. 1, Halver, J. E. and Tiews, K., Eds., Heenemann, Berlin, 1979, 127.
11. **Furukawa, A.**, Diet of yellowtail: its digestibility, *Suisan Zoshoku* Extra No. 6, 51, 1966.
12. **Furuichi, M. and Yone, Y.**, Effect of dietary dextrin levels on the growth and feed efficiency, the chemical composition of liver and dorsal muscle, and the absorption of dietary protein and dextrin in fishes, *Nippon Suisan Gakkaishi*, 46, 225, 1980.
13. **Kanbe, K.**, Study on Digestion Process of Protein in Young Yellowtail, M.S. thesis, Kochi University, Nankoku, Japan, 1976.
14. **Takeda, M., Shimeno, S., Hosokawa, H., Kajiyama, H., and Kaisho, K.**, The effect of dietary calorie-to-protein ratios on the growth, feed conversion and body composition of young yellowtail, *Nippon Suisan Gakkaishi*, 41, 443, 1975.
15. **Deshimaru, O. and Kuroki, K.**, Studies on the Optimum Levels of Protein and Lipid in Yellowtail Diets, Reports of Kagoshima Prefectural Fishery Experimental Station, 44-79, 1983.
16. **Kuwabara, H., Yamaguchi, M., Hagino, S., Takeda, M., Shimeno, S., Kajiyama, H., Kubota, S., and Miyazaki, T.**, Sources of fish meal and trace nutrients in formulated diet of yellowtail-I. Growth and feed conversion, in Abstracts of Oral Presentation at the Autumn Meeting of Nippon Suisan Gakkai at Fukuyama, 1982, 105.
17. **Shimeno, S., Hosokawa, H., Takeda, M., and Ono, T.**, Sources of fish meal and trace nutrients in formulated diet of yellowtail-II. General components and trace nutrients in the fish, in Abstracts of Oral Presentation at the Autumn Meeting of Nippon Suisan Gakkai at Fukuyama, 1982, 105.
18. **Takii, K., Shimeno, S., Nakamura, M., Itoh, Y., Obatake, A., Kumai, H., and Takeda, M.**, Evaluation of soy protein concentrate as a partial substitute for fish meal protein in practical diet for yellowtail, in *The Current Status of Fish Nutrition in Aquaculture*, Takeda, M. and Watanabe, T., Eds., Tokyo University of Fisheries, Tokyo, Japan, 1990, 281.
19. **Shimeno, S., Hosokawa, H., Takeda, M., Kajiyama, H., and Kaisho, T.**, Effect of dietary lipid and carbohydrate on growth, feed conversion and body composition in young yellowtail, *Nippon Suisan Gakkaishi*, 51, 1893. 1985.
20. **Shimeno, S., Sasaki, H., Takeda, M., and Kajiyama, H.**, Effect of dietary composition on the growth, feed conversion and body composition in young yellowtail: lipid and carbohydrate levels in diet, in Abstracts of Oral Presentation at the Spring Meeting of Nippon Suisan Gakkai at Tokyo, 1979, 124.
21. **Furuichi, M.**, Studies on the Utilization of Carbohydrate by Fishes, Report of Fishery Research Laboratory, Kyushu University, No. 6, 1, 1983.
22. **Furuichi, M., Taira, H., and Yone, Y.**, Availability of carbohydrate in nutrition of yellowtail, *Nippon Suisan Gakkaishi*, 52, 99, 1986.
23. **Shimeno, S., Hosokawa, H., Kajiyama, H., and Takeda, M.**, Effect of Dietary Carbohydrates on the Growth, Feed Conversion, Body Composition and Liver Enzymes of Young Yellowtail, Reports of Fisheries Laboratory, Kochi University, No. 3, 89, 1978.
24. **Deshimaru, O., Kuroki, K., and Shintani, K.**, Studies on the Nutrition of Lipid and Protein in Yellowtail Diet, Reports of Kagoshima Prefectural Fishery Experimental Station, 1-48, 1986.
25. **Takeda, M.**, Nutrition and diet in yellowtail-3. Nutrition of lipids, *Fish Culture*, 22(5), 98, 1985.
26. **Deshimaru, O., Kuroki, K., and Yone, Y.**, Nutritive values of various oils for yellowtail, *Nippon Suisan Gakkaishi*, 48, 1155, 1982.
27. **Tsukahara, H., Furukawa, A., and Funae, K.**, Studies on feed for fish-VII. The effects of dietary fat on the growth of yellowtail, *Bull. Naikai Regional Fish. Res. Lab.*, No. 24, 29, 1967.
28. **Yone, Y.**, Essential fatty acid and nutritive value of lipids, in *Fish Culture and Dietary Lipid*, Nippon Suisan Gakkai, Ed., Koseisha Koseikaku, Tokyo, 1978, 23.
29. **Shimeno, S., Hosokawa, H., Takeda, M., Takayama, S., Fukui, A., and Sasaki, H.**, Adaptation of hepatic enzymes to dietary lipid in young yellowtail, *Nippon Suisan Gakkaishi*, 47, 63, 1981.
30. **Sakaguchi, H. and Hamaguti, A.**, Influence of oxidized oil and vitamin A on the culture of yellowtail, *Nippon Suisan Gakkaishi*, 35, 1207, 1969.

31. **Kubota, S., Funahashi, N., Endo, M., and Miyazaki, T.**, Studies on nutritional myopathy in cultured fishes-I. Nutritional myopathy in cultured yellowtail, *Fish Pathol.*, 15, 75, 1980.
32. **Murai, T., Akiyama, T., Ogata, H., and Suzuki, T.**, Interaction of dietary oxidized fish oil and glutathione on fingerling yellowtail, *Nippon Suisan Gakkaishi*, 54, 145, 1988.
33. **Hosokawa, H., Teraoka, R., Saito, Y., and Takeda, M.**, Qualitative requirements of water-soluble vitamins in yellowtail, in Abstracts of Oral Presentation at the Spring Meeting of Nippon Suisan Gakkai at Tokyo, 1979, 127.
34. **Teraoka, R.**, Study on Qualitative Requirements of Water-Soluble Vitamins in Yellowtail, M.S. thesis, Kochi University, Nankoku, Japan, 1980.
35. **Hosokawa, H., Ueno, S., Hirata, S., and Takeda, M.**, Quantitative requirements of water-soluble vitamins in yellowtail-III and IV. Thiamin, biotin, folic acid, cyanocobalamin, inositol and ascorbic acid, in Abstracts of Oral Presentation at the Spring Meeting of Nippon Suisan Gakkai at Tokyo, 1982, 102.
36. **Ueno, S.**, Study on Quantitative Requirements of Water-Soluble Vitamins in Yellowtail, M.S. thesis, Kochi University, Nankoku, Japan, 1983.
37. **Toyoda, Y.**, Study on Quantitative Requirements of Fat-Soluble Vitamins in Yellowtail, M.S. thesis, Kochi University, Nankoku, Japan, 1985.
38. **Hosokawa, H., Ueno, S., Toyoda, Y., and Takeda, M.**, Quantitative requirements of water-soluble vitamins in yellowtail-V. Effect of the vitamin mixture based on all the quantitative vitamin requirements, in Abstracts of Oral Presentation at the Spring Meeting of Nippon Suisan Gakkai at Tokyo, 1983, 44.
39. **Teishima, H.**, Study on Immunopotentiation of Supplemental Vitamins in Yellowtail, M.S. thesis, Kochi University, Nankoku, Japan, 1990.
40. **Hosokawa, H., Ueno, S., and Takeda, M.**, Quantitative requirements of water-soluble vitamins in yellowtail-VI. Effect of dietary protein levels on pyridoxine requirement, in Abstracts of Oral Presentation at the Autumn Meeting of Nippon Suisan Gakkai at Kyoto, 1983, 89.
41. **Hosokawa, H., Toyoda, Y., and Takeda, M.**, Effect of dietary lipid levels on vitamin E requirement in yellowtail, in Abstracts of Oral Presentation at the Spring Meeting of Nippon Suisan Gakkai at Tokyo, 1985, 54.
42. **Hosokawa, H. and Takeda, M.**, Assessment of the thiamin status in yellowtail, in Abstracts of Papers of Satellite Symposium at the XIII Int. Congress of Nutrition in Brighton, 1985, 1.
43. **Shimizu, T.**, Minerals, in *Fish Nutrition and Diets*, Yone, Y., Ed., Koseisha Koseikaku, Tokyo, 1985, 54.
44. **Halver, J. E.**, Nutrition of salmonoid fishes-III. Water soluble vitamin requirements of chinook salmon, *J. Nutr.*, 62, 225, 1957.
45. **Hosokawa, H., Takeda, M., Tachi, T., and Hayami, H.**, Study on a basal diet for yellowtail-II. Mineral mixture, in Abstracts of Oral Presentation at the Autumn Meeting of Nippon Suisan Gakkai at Kyoto, 1974, 16.
46. **Makino, H.**, Study on Mineral Requirements in Yellowtail, M.S. thesis, Kochi University, Nankoku, Japan, 1990.
47. **Ikeda, Y.**, Diagnostic Studies on Hematological and Chemical Constituents of Blood in Cultured Yellowtail, Ph.D. thesis, Kyoto University, Kyoto, Japan, 1976.
48. **Kubota, S., Kamiya, N., Takeda, M., Shimeno, S., and Kuwabara, H.**, Pathological and histological study on young yellowtail fed various ferrous compounds, in Abstracts of Oral Presentation at the Meeting of Nippon Gyobyo Gakkai at Tokyo, 1983, 6.
49. **Kuwabara, H., Takeda, M., Shimeno, S., Hosokawa, H., Kubota, S., Kamiya, N., and Kajiyama, H.**, Effect of adding ferrous compounds to yellowtail diets, in Abstracts of Oral Presentation at the Spring Meeting of Nippon Suisan Gakkai at Tokyo, 1983, 43.
50. **Takeda, M.**, Nutrition and diet in yellowtail-8. Present and future formulated diets, *Fish Culture*, 22(10), 102, 1985.
51. **Kanazawa, A.**, Microparticulate diets, in *Fish Nutrition and Diets*, Yone, Y., Ed., Koseisha Koseikaku, Tokyo, 1985, 99.

INDEX

Milton Keynes UK
Ingram Content Group UK Ltd.
UKHW051934141024
449569UK00027B/1490

9 781138 560000